Standard Grade | General

Physics

Leckie × Leckie

First exam published in 2004.
Published by Leckie & Leckie Ltd, 3rd Floor, 4 Queen Street, Edinburgh EH2 1JE
tel: 0131 220 6831 fax: 0131 225 9987 enquiries@leckieandleckie.co.uk www.leckieandleckie.co.uk

ISBN 978-1-84372-641-8

A CIP Catalogue record for this book is available from the British Library.

Leckie & Leckie is a division of Huveaux plc.

Leckie & Leckie is grateful to the copyright holders, as credited at the back of the book, for permission to use their material. Every effort has been made to trace the copyright holders and to obtain their permission for the use of copyright material. Leckie & Leckie will gladly receive information enabling them to rectify any error or omission in subsequent editions.

[BLANK PAGE]

FOR OFFICIAL USE

G

K & U	PS

Total Marks

3220/401

NATIONAL QUALIFICATIONS 2004

FRIDAY 28 MAY
9.00 AM – 10.30 AM

PHYSICS
STANDARD GRADE
General Level

Fill in these boxes and read what is printed below.

Full name of centre

Town

Forename(s)

Surname

Date of birth
Day Month Year

Scottish candidate number

Number of seat

1 All questions should be answered.

2 The questions may be answered in any order but all answers must be written clearly and legibly in this book.

3 For questions 1–5, write down, in the space provided, the letter corresponding to the answer you think is correct. There is only **one** correct answer.

4 For questions 6–18, write your answer where indicated by the question or in the space provided after the question.

5 If you change your mind about your answer you may score it out and replace it in the space provided at the end of the answer book.

6 Before leaving the examination room you must give this book to the invigilator. If you do not, you may lose all the marks for this paper.

DO NOT WRITE IN THIS MARGIN

K&U | PS

Marks

1. The diagram shows a ray of light reflected from a plane mirror.

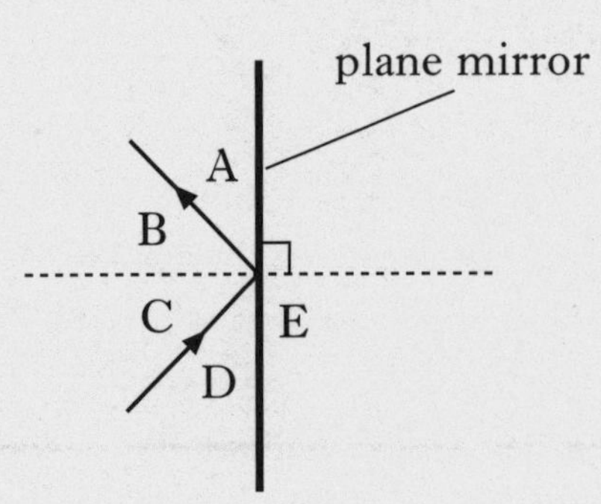

Which of the labelled angles is the angle of reflection?

Answer ☐ **1**

2. The instrument used to measure the temperature of a human body is called

A an endoscope

B an ohmmeter

C an oscilloscope

D a stethoscope

E a thermometer.

Answer ☐ **1**

3. Which one of the following radiations is used in a laser?

A Gamma rays

B X-rays

C Visible light

D Microwaves

E Radio waves

Answer ☐ **1**

DO NOT WRITE IN THIS MARGIN

K&U	PS

Marks

4. A newton balance is used to measure

A distance

B force

C gravitational potential energy

D kinetic energy

E power.

Answer ☐ **1**

5. The International Space Station has an orbital height of 352 kilometres and a period of 92 minutes.

A geostationary satellite has an orbital height of 35 900 kilometres and a period of 1440 minutes.

Which of the following gives the orbital height of a satellite that has a period of 102 minutes?

A 144 kilometres

B 352 kilometres

C 833 kilometres

D 35 900 kilometres

E 44 100 kilometres

Answer ☐ **1**

[Turn over

DO NOT WRITE IN THIS MARGIN

K&U PS

Marks

6. A worker in a telephone call centre wears a headset that has an earphone and a microphone.

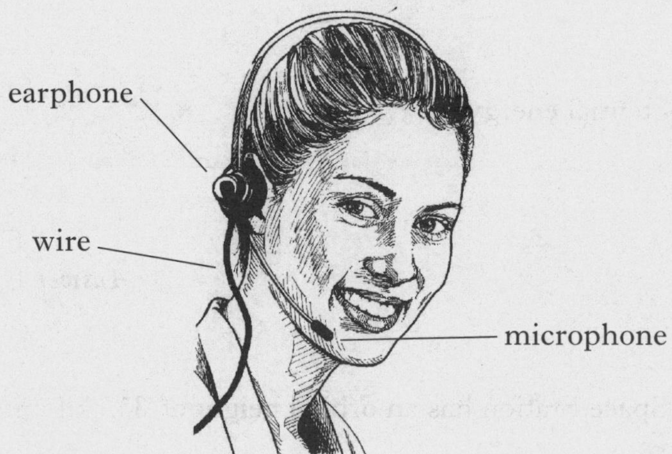

(*a*) State the energy transformation that takes place in the

(i) earphone

....

(ii) microphone.

.... **2**

(*b*) In the passage below, circle **one** phrase in each set of brackets to make the statement correct.

{An electrical signal / A light beam / A sound wave} is transmitted along the wire

at a speed {less than / equal to / greater than} the speed of sound. **2**

DO NOT WRITE IN THIS MARGIN

K&U	PS

Marks

7. A girl is at a fireworks display.

(*a*) The girl notices that when a firework explodes high in the air, she sees the flash before she hears the bang from the firework.

Explain why there is a delay between seeing the flash and hearing the bang.

....

....

.... **1**

(*b*) The girl wants to calculate how far away the firework is when it explodes. She uses a stopwatch to measure the time interval between the flash and the bang from the firework. The reading on the stopwatch is 0·8 second.

(i) What additional information is needed to calculate this distance?

.... **1**

(ii) Explain why the distance calculated by the girl is likely to be inaccurate.

....

....

.... **2**

[Turn over

DO NOT WRITE IN THIS MARGIN

K&U	PS

Marks

8. A model car contains an electric motor, operated by a 6 volt battery. The speed of the motor is adjusted by a hand-held control. The hand-held control contains a variable resistor.

The circuit is shown below.

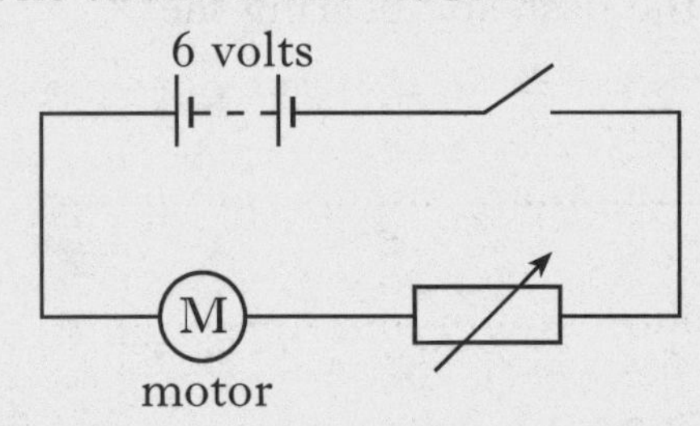

(*a*) When the resistance of the variable resistor is set to 8 ohms, the voltage across the variable resistor is 2 volts.

(i) Calculate the current in the variable resistor.

Space for working and answer

2

(ii) Calculate the voltage across the motor at this setting of the variable resistor.

Space for working and answer

1

(*b*) The resistance of the variable resistor is decreased.

Explain what happens to the speed of the motor.

....

....

....

.... **2**

(*c*) Give one **other** use for a variable resistor.

.... **1**

DO NOT WRITE IN THIS MARGIN

K&U | PS

Marks

9. A food mixer is used to prepare food.

(*a*) The double insulation symbol is displayed on the food mixer.

(i) Draw the double insulation symbol.

Space for symbol

1

(ii) Which wire is **not** needed in the flex of the food mixer?

.... **1**

(*b*) The diagram below shows the main parts of the electric motor used in the food mixer.

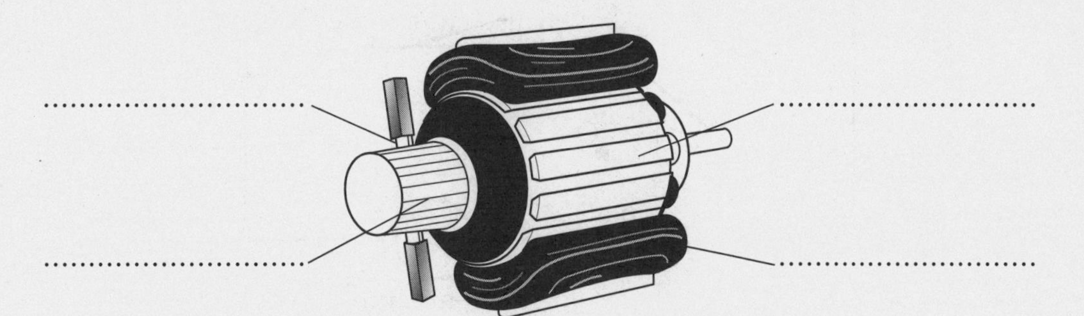

Label the diagram using the parts of the motor listed below.

Field coil (magnet) **Brush** **Rotating coil** **Commutator** **2**

[Turn over

DO NOT WRITE IN THIS MARGIN

K&U	PS

Marks

10. A camping light contains an 8 watt discharge tube, an 8 watt filament lamp and a 12 volt battery.

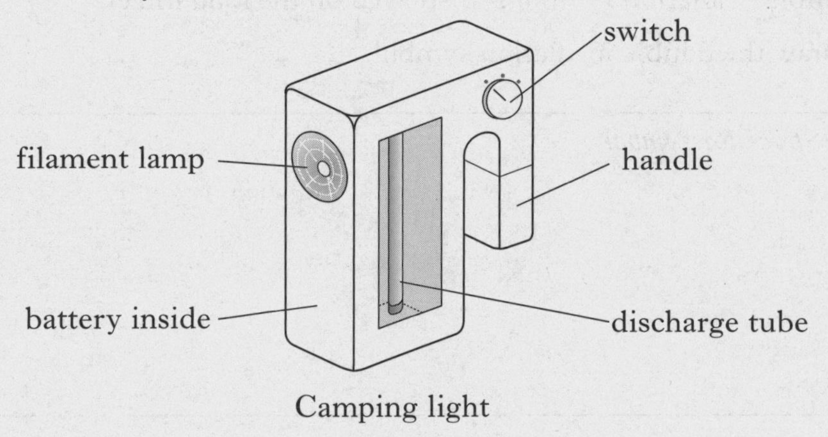

Camping light

The circuit diagram for the camping light is shown.

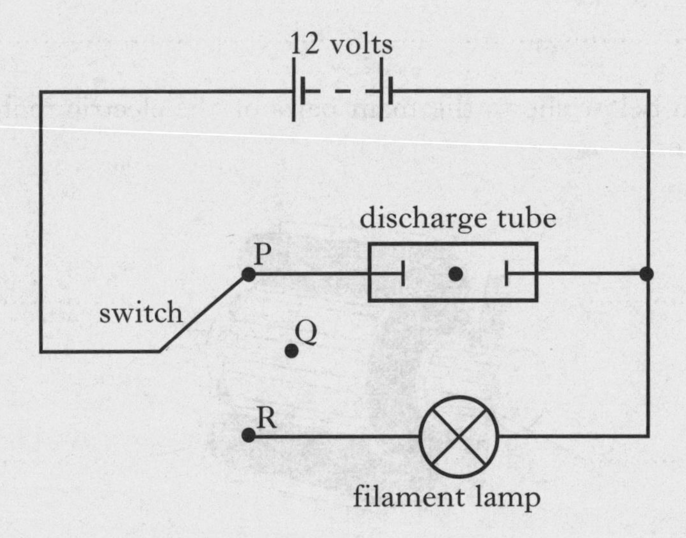

Circle the correct answer for each of the questions about the camping light.

(*a*) The camping light is **off** when the switch is in position {P / Q / R}. **1**

(*b*) The operating voltage of the filament lamp is {6 / 8 / 12} volts. **1**

DO NOT WRITE IN THIS MARGIN

K&U	PS

Marks

10. (continued)

(*c*) The filament lamp and the discharge tube are constructed as shown below.

Filament lamp

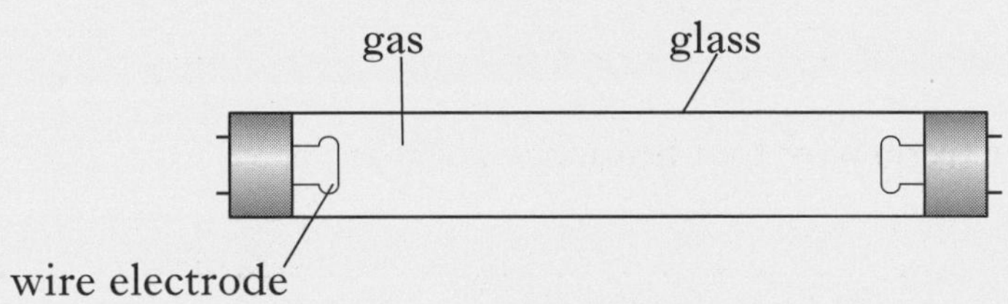

Discharge tube

(i) The useful energy transformation in the filament lamp takes place

in the {glass / wire / gas}. **1**

(ii) The useful energy transformation in the discharge tube takes place

in the {glass / wire / gas}. **1**

(iii) The electrical energy transformed each second by the discharge

tube is {smaller than / the same as / greater than} the electrical energy transformed each second by the filament lamp. **1**

(iv) The heat energy produced each second by the discharge tube is

{smaller than / the same as / greater than} the heat energy produced each second by the filament lamp. **1**

DO NOT WRITE IN THIS MARGIN

K&U | PS

Marks

11. Read the following passage about sound.

Sound with a frequency below 20 hertz is called infrasound. Sound with a frequency above the range of human hearing is called ultrasound.

Elephants communicate using infrasound. Elephants can detect low level infrasound through their feet.

Bats use ultrasound to navigate. They send out ultrasound pulses that reflect off objects. The bats note how long it takes the pulses to return.

Ultrasound is also used in medicine.

(*a*) Suggest a frequency that could be detected by an elephant through its feet.

.... **1**

(*b*) State the highest frequency that humans can hear.

.... **1**

(*c*) State the unit of sound level.

.... **1**

(*d*) A bat sends out an ultrasound pulse of frequency 45 000 hertz. The pulse is reflected and returns to the bat after 0·2 second.

Calculate the total distance that the pulse travels.
[The speed of sound in air is 340 metres per second.]

Space for working and answer

2

(*e*) Give an example of a use of ultrasound in medicine.

....

.... **1**

DO NOT WRITE IN THIS MARGIN

K&U	PS

Marks

12. Two students are revising for a Physics test.

(*a*) One student draws a simple model of an atom.

Complete the diagram by adding the following labels.

protons **neutrons** **electrons**

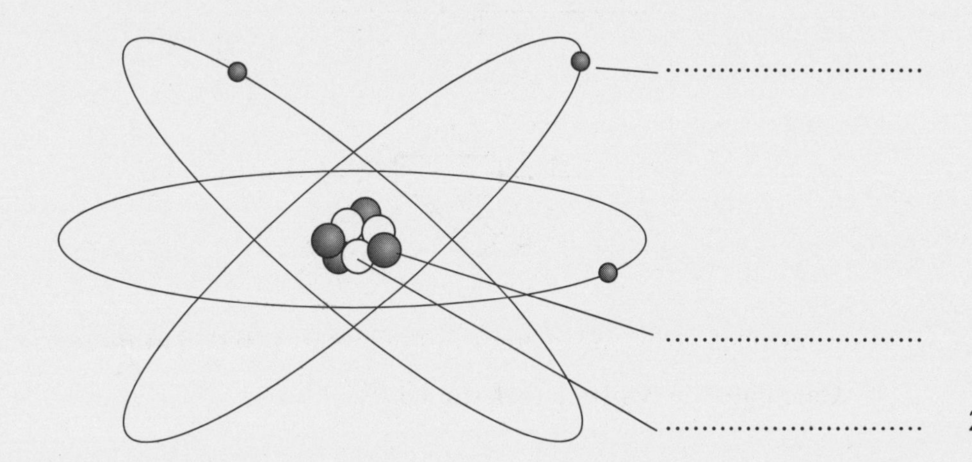

2

(*b*) The other student writes incomplete statements about nuclear radiation.

Complete the statements using words from the following list.

alpha **beta** **gamma** **becquerels** **sieverts**

(i) The radiation that has the greatest range is 1

(ii) The radiation that is absorbed by a sheet of paper is....................... 1

(iii) Dose equivalent is measured in 1

(*c*) The students ask each other about nuclear radiation and safety.
State **two** safety precautions necessary when handling radioactive substances.

Precaution 1..

....

Precaution 2..

.... 2

[Turn over

DO NOT WRITE IN THIS MARGIN

K&U PS

Marks

13. A company makes sunglasses. The company uses a light meter to measure how much light passes through different types of glass. The light meter contains an ammeter, an LDR and a 6 volt battery as shown.

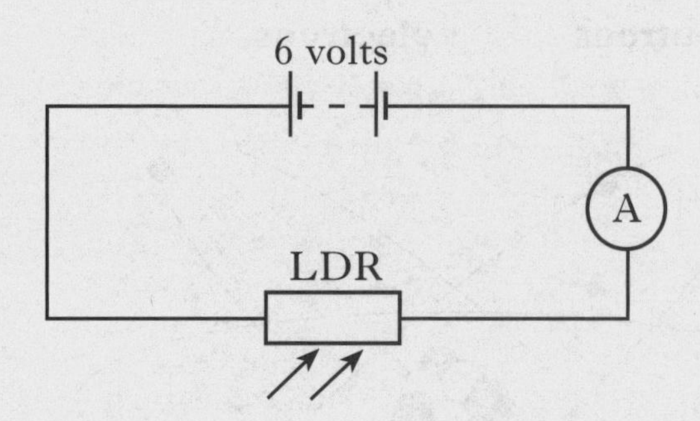

(*a*) For one type of glass, the current in the circuit is 0·005 ampere.

(i) Calculate the resistance of the LDR.

Space for working and answer

2

(ii) The intensity of the light shining on the LDR is increased.

(A) State what happens to the resistance of the LDR.

.... 1

(B) State what happens to the current in the circuit.

.... 1

DO NOT WRITE IN THIS MARGIN

K&U	PS

Marks

13. (continued)

(*b*) Three types of glass, X, Y and Z are tested as shown below to find out how much light passes through each. The same source of light is used in all three tests.

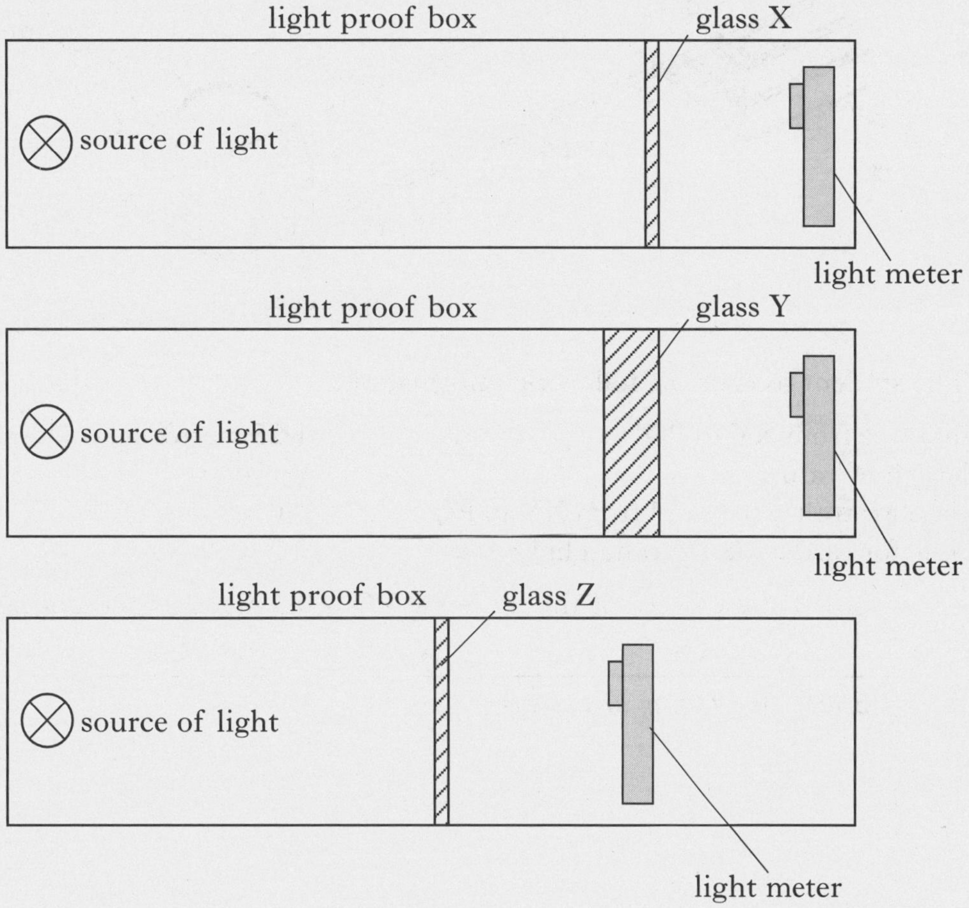

light proof box
glass X
source of light
light meter

light proof box
glass Y
source of light
light meter

light proof box
glass Z
source of light
light meter

Give **two** reasons why this is not a fair test.

Reason 1 ....

....

Reason 2 ....

.... **2**

[Turn over

DO NOT WRITE IN THIS MARGIN

K&U	PS

Marks

14. A student releases a trolley from rest at XY near the top of a slope. The trolley moves down the slope. A card attached to the trolley passes through a light gate at PQ near the bottom of the slope.

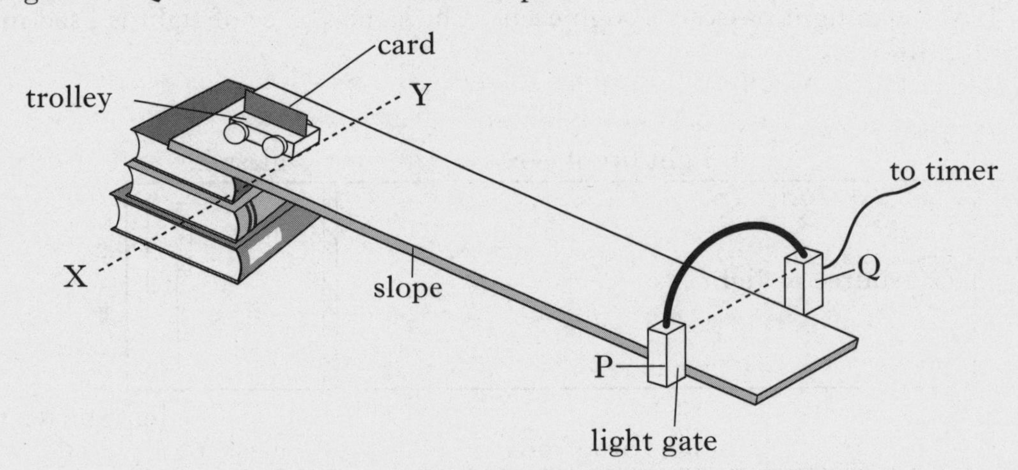

(*a*) The student records the following information.

distance from XY to PQ	1·25 metres
length of card	0·1 metre
time for trolley to travel from XY to PQ	5·0 seconds
time for card to pass through light gate	0·2 second.

(i) Show that the speed of the trolley **at PQ** is 0·5 metre per second.

Space for working and answer

2

(ii) Calculate the acceleration of the trolley down the slope.

Space for working and answer

2

DO NOT WRITE IN THIS MARGIN

K&U	PS

Marks

14. (a) (continued)

(iii) Draw a speed-time graph for the motion of the trolley from when it is released at XY until it passes through PQ.

Units and numerical values **must** be shown on both axes.

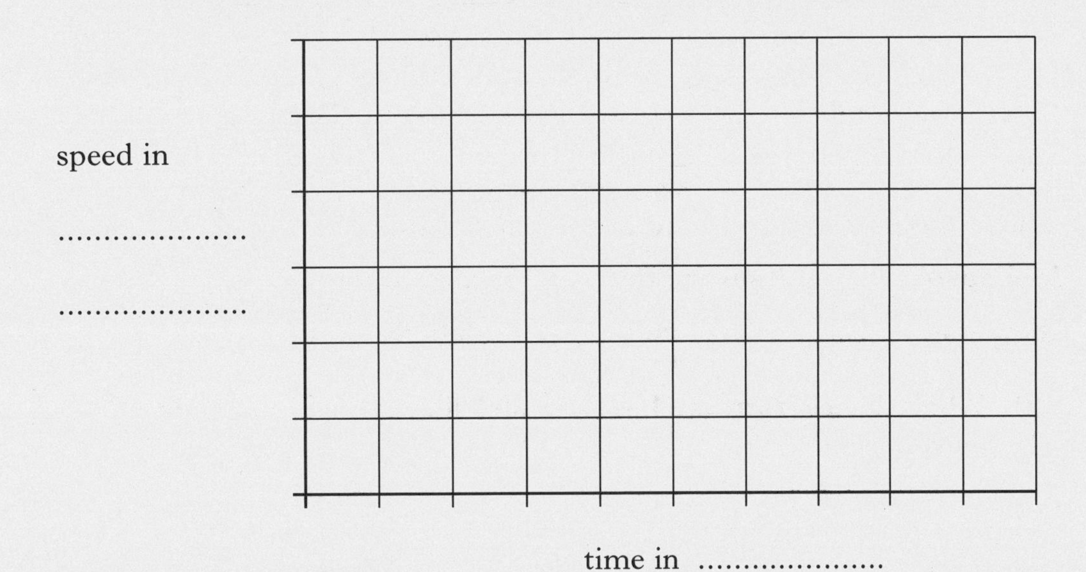

3

(b) The student moves the light gate up the slope to RS, as shown, and repeats the experiment.

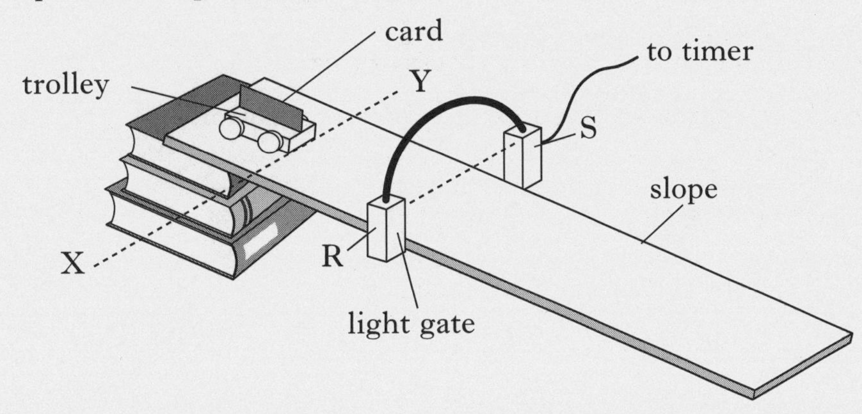

No other changes are made to the apparatus. The trolley is again released from rest at XY.

Complete the sentences below by circling the correct answers.

Compared to the first part of the experiment:

(i) the speed of the trolley at the light gate is {less / the same / greater} **1**

(ii) the acceleration of the trolley down the slope is {less / the same / greater}. **1**

DO NOT WRITE IN THIS MARGIN

K&U PS

Marks

15. A student designs an electronic system that produces a sound when the temperature in a fish tank falls below a certain value.

(*a*) A block diagram of the system is shown.

Complete the block diagram by filling in the two missing labels.

[ ] → Process → [ ] **1**

(*b*) The following components are available to the student.

loudspeaker	**solenoid**	**microphone**
thermistor	**solar cell**	**electric motor**

(i) Which device **from the list** is suitable for sensing a change in temperature?

.... **1**

(ii) Which device **from the list** is suitable for producing a sound?

.... **1**

(*c*) The student uses a transistor as the process device.

In the space below, draw the circuit symbol for a transistor.

Space for circuit symbol **1**

DO NOT WRITE IN THIS MARGIN

K&U	PS

Marks

16. A farmer installs a wind-powered generator on a farm.

(*a*) On a particular day, the generator has a constant power output of 20 kilowatts.

(i) How many kilowatt-hours of electrical energy are generated in 8 hours?

Space for working and answer

2

(ii) Electrical energy from the local supply company costs the farmer 9 pence per kilowatt-hour.

Calculate how much money the farmer saves by using the generator for 8 hours.

Space for working and answer

2

(*b*) Wind is a renewable source of energy.

(i) Name one other renewable source of energy.

.... **1**

(ii) Name one non-renewable source of energy.

.... **1**

DO NOT WRITE IN THIS MARGIN

K&U	PS

Marks

17. The diagram shows how electricity is distributed from a power station to different consumers by the National Grid.

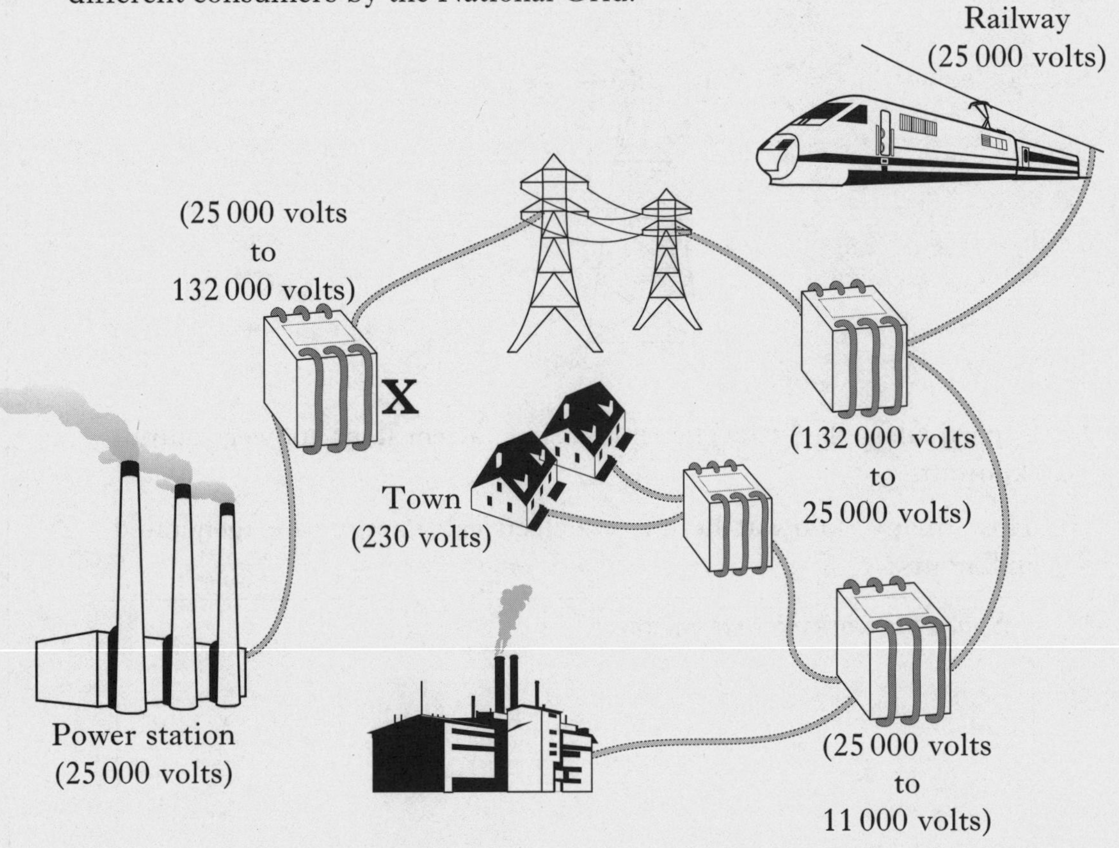

(*a*) Name the part labelled X in the diagram and state its purpose.

....

....

.... **2**

(*b*) Why are high voltages used in the transmission of electrical energy?

....

....

.... **1**

DO NOT WRITE IN THIS MARGIN

K&U	PS

Marks

17. (continued)

(*c*) State the voltage at which electrical energy is used by the railway.

.... **1**

(*d*) Calculate the ratio $\frac{\text{number of turns in primary}}{\text{number of turns in secondary}}$ in the transformer used to supply energy for the railway.

Space for working and answer

2

[Turn over

DO NOT WRITE IN THIS MARGIN

K&U	PS

Marks

18. A boy is interested in astronomy.

(*a*) The boy writes his address in the Universe.

Complete the address given below by writing, **in the correct order**, the missing lines using terms from the following list.

Earth **Milky Way (our galaxy)** **Solar System**

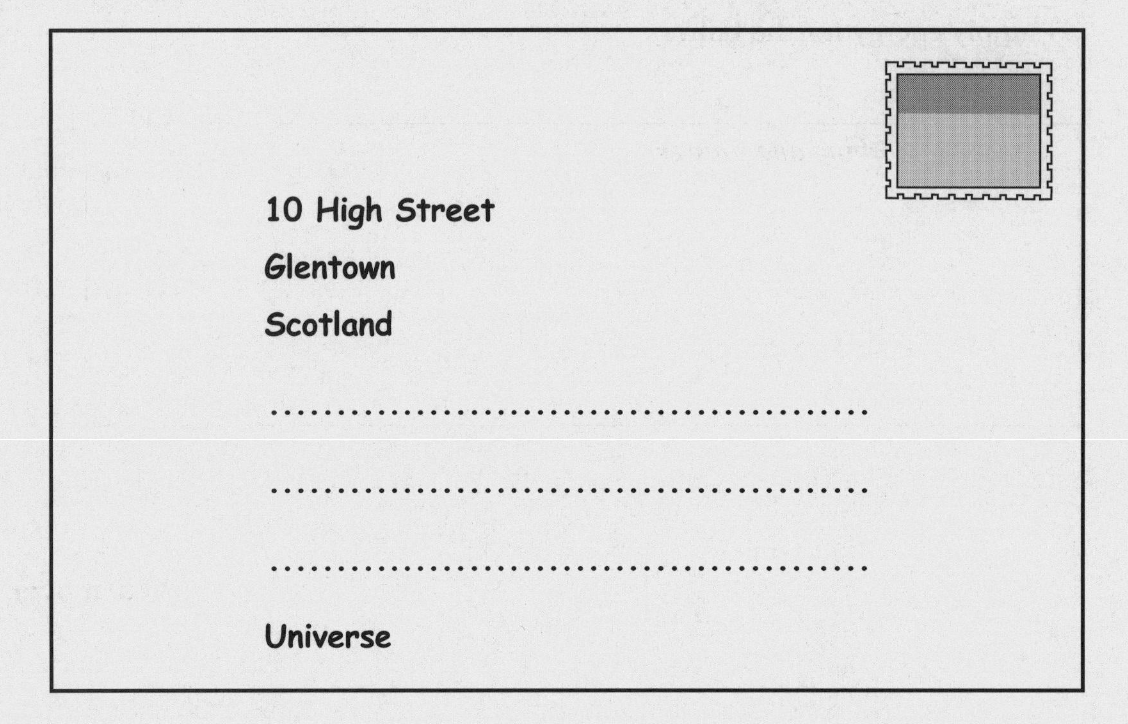

2

(*b*) The boy reads the following passage in an astronomy book.

You can view planets, moons and stars using a telescope.
Jupiter, the largest planet in our Solar System, takes 12 Earth years to orbit the Sun. The largest of Jupiter's many moons is called Ganymede.
Sirius, also known as the dog star, is the brightest star in the sky, apart from the Sun.

(i) Name one astronomical object, **mentioned in the passage**, that can only be seen by reflected light.

.... 1

(ii) Name one astronomical object, **mentioned in the passage**, that generates light.

.... 1

(iii) Which object, **mentioned in the passage**, is furthest away from Earth?

.... 1

DO NOT WRITE IN THIS MARGIN

K&U	PS

Marks

18. (*b*) (continued)

(iv) Complete the diagram of a telescope below, by naming the two lenses.

eye

.... lens .... lens **2**

[*END OF QUESTION PAPER*]

DO NOT WRITE IN THIS MARGIN

K&U	PS

YOU MAY USE THE SPACE ON THIS PAGE TO REWRITE ANY ANSWER YOU HAVE DECIDED TO CHANGE IN THE MAIN PART OF THE ANSWER BOOKLET. TAKE CARE TO WRITE IN CAREFULLY THE APPROPRIATE QUESTION NUMBER.

2005 | General

[BLANK PAGE]

FOR OFFICIAL USE

G

K & U	PS

Total Marks

3220/401

NATIONAL QUALIFICATIONS 2005

TUESDAY, 24 MAY
9.00 AM – 10.30 AM

PHYSICS
STANDARD GRADE
General Level

Fill in these boxes and read what is printed below.

Full name of centre

Town

Forename(s)

Surname

Date of birth
Day Month Year

Scottish candidate number

Number of seat

1 All questions should be answered.

2 The questions may be answered in any order but all answers must be written clearly and legibly in this book.

3 For questions 1–6, write down, in the space provided, the letter corresponding to the answer you think is correct. There is only **one** correct answer.

4 For questions 7–21, write your answer where indicated by the question or in the space provided after the question.

5 If you change your mind about your answer you may score it out and replace it in the space provided at the end of the answer book.

6 Before leaving the examination room you must give this book to the invigilator. If you do not, you may lose all the marks for this paper.

SAB 3220/401 6/23870

DO NOT WRITE IN THIS MARGIN

K&U PS

Marks

1. Which of the following is the circuit symbol for a fuse?

A 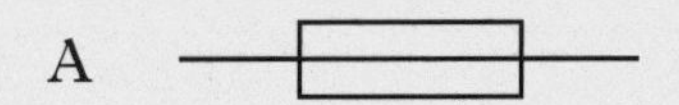

B 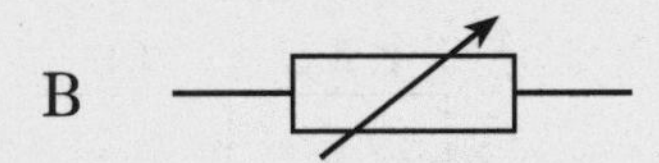

C

D

E

Answer ☐ **1**

2. What is the main energy transformation that takes place in a thermocouple?

A Heat to light

B Electrical to heat

C Heat to electrical

D Light to heat

E Heat to chemical

Answer ☐ **1**

3. A 20 newton weight is hung on a spring balance. The spring extends by 0·10 metre. The weight is removed and a bag of potatoes is hung on the balance. The spring extends by 0·15 metre.

What is the weight of the bag of potatoes?

A 10 newtons

B 15 newtons

C 20 newtons

D 30 newtons

E 50 newtons

Answer ☐ **1**

DO NOT WRITE IN THIS MARGIN

K&U PS

Marks

4. A car designer wants to increase the maximum acceleration of a car.

Which entry shows what should be done to the engine force and the mass of the car?

	Engine force	*Mass*
A	keep the same	increase
B	increase	decrease
C	increase	keep the same
D	decrease	increase
E	decrease	keep the same

Answer ☐ **1**

5. The diagrams below show the forces acting on a number of moving objects.

Which object is moving at constant speed?

A
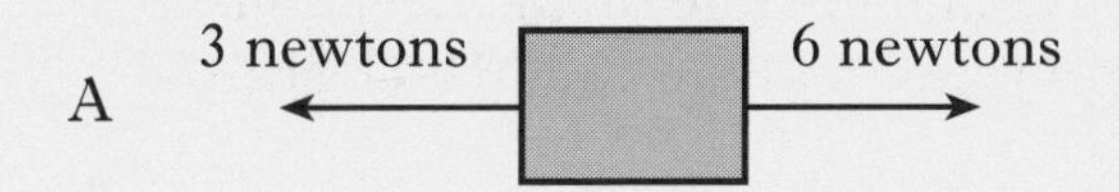

B
3 newtons
6 newtons
6 newtons

C
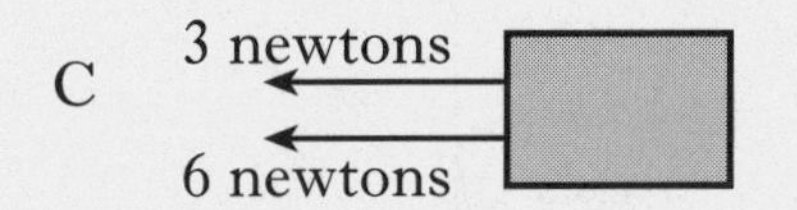

D
6 newtons
6 newtons
3 newtons
3 newtons

E
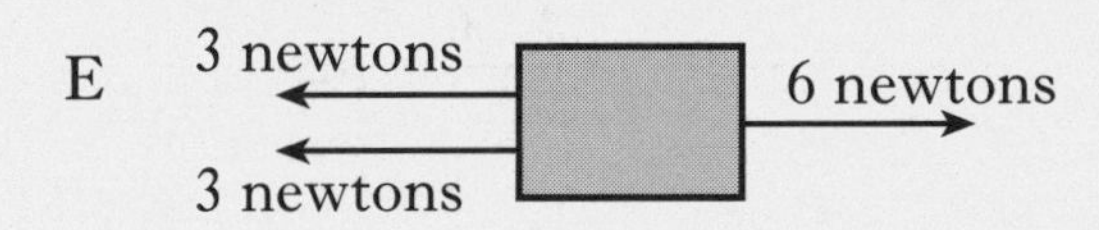

Answer ☐ **1**

6. Which row gives the correct units for work done, energy and power?

	Work done	*Energy*	*Power*
A	newton	joule	watt
B	joule	joule	watt
C	newton	watt	joule
D	watt	newton	watt
E	joule	watt	newton

Answer ☐ **1**

DO NOT WRITE IN THIS MARGIN

K&U | PS

Marks

7. Two students are investigating a telephone system in a laboratory.

(*a*) An oscilloscope is connected to the microphone in one of the telephones. One student whistles several times into this microphone and the electrical signals shown are obtained.

A

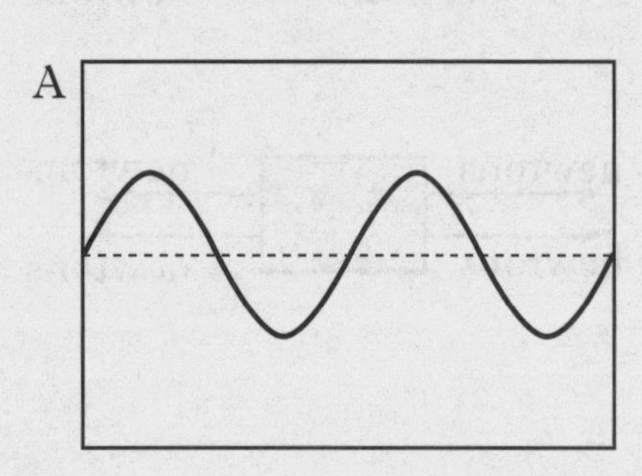

B

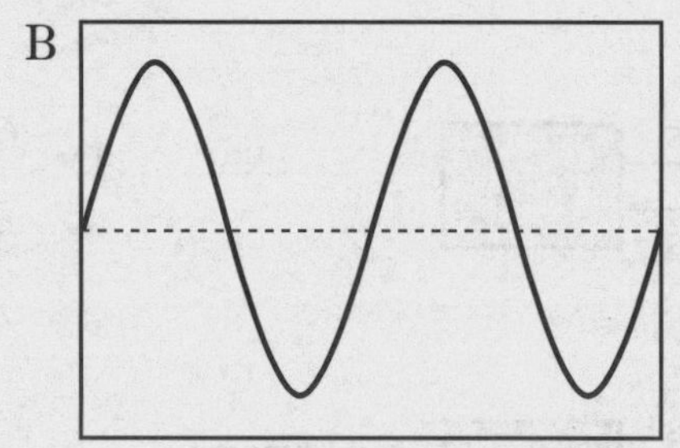

C

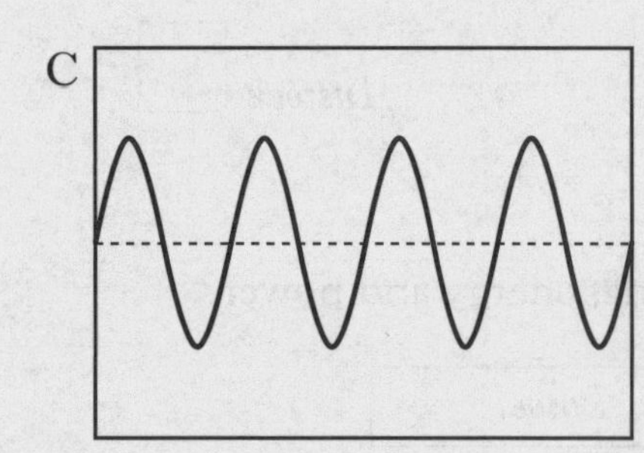

D

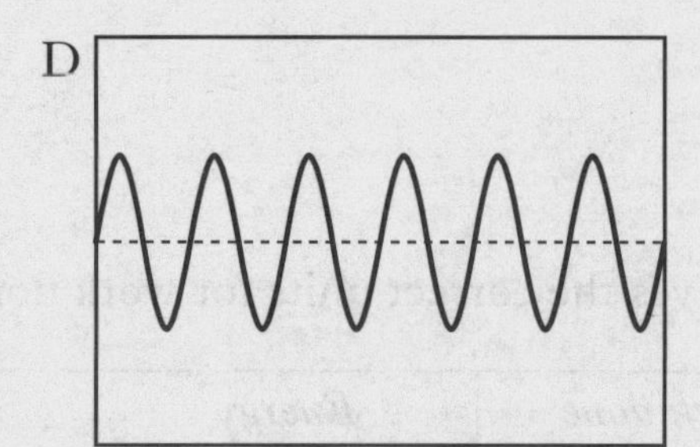

All the traces shown are obtained without changing the controls on the oscilloscope.

Which of these electrical signals is caused by

(i) the highest frequency sound *Answer* ☐

(ii) the loudest sound? *Answer* ☐ **2**

DO NOT WRITE IN THIS MARGIN

K&U	PS

Marks

7. (continued)

(*b*) In the telephone system, electrical signals carry the information from the transmitter to the receiver.

One student makes a loud sound. The other student hears this sound through the telephone and also directly through the air.

Explain which sound reaches the student first.

....

....

....

.... **2**

[Turn over

DO NOT WRITE IN THIS MARGIN

K&U	PS

Marks

8. In a research laboratory, water waves are generated in a tank. During one test the wave shown travels along the tank at 2·5 metres per second.

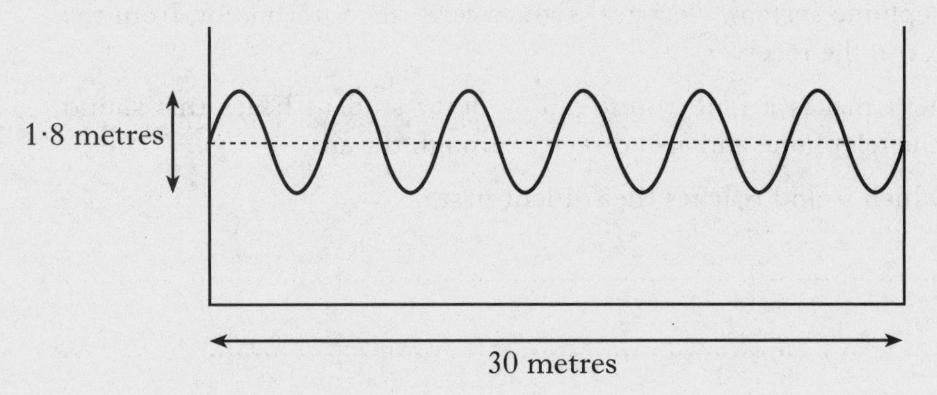

(*a*) Calculate the amplitude of the wave shown.

Space for working and answer

1

(*b*) Calculate the wavelength of the wave shown.

Space for working and answer

2

(*c*) Calculate the frequency of the wave shown.

Space for working and answer

2

DO NOT WRITE IN THIS MARGIN

K&U	PS

Marks

9. A two-speed hot air blower is used in a factory. The blower operates from a 110 volt supply. The blower contains a heater, and a fan attached to a motor. The blower is switched on by closing switch S_1.

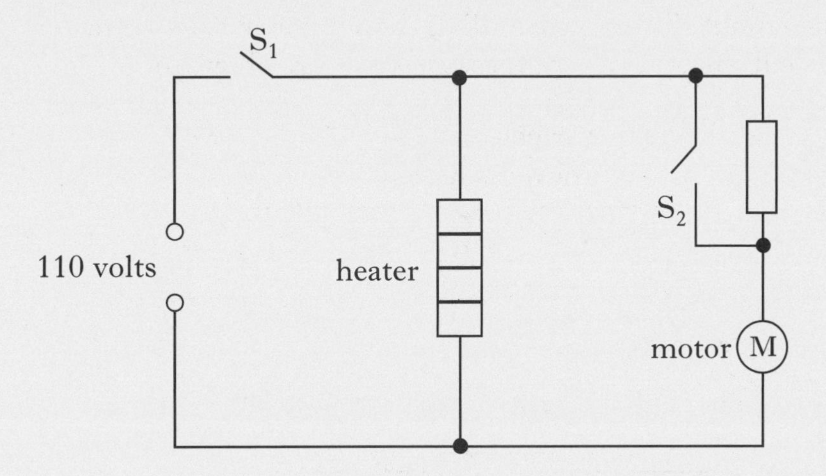

(*a*) What is the voltage across the heater when the blower is operating?

.... **1**

(*b*) Explain why switch S_2 should now be closed for the blower to operate at high speed.

....

.... **2**

(*c*) When operated at high speed, the blower is rated at 2000 watts. The blower is operated at high speed for 8 hours.

(i) Calculate the number of kilowatt-hours of energy it uses in this time.

Space for working and answer

2

(ii) Electricity costs 9 pence per kilowatt-hour.
Calculate the cost of operating the blower for 8 hours.

Space for working and answer

2

DO NOT WRITE IN THIS MARGIN

K&U	PS

Marks

10. A variable power supply, an ammeter and a voltmeter are used to investigate how the current in a thermistor changes as the voltage across the thermistor changes.

(*a*) Complete the circuit diagram, including the ammeter and voltmeter, to show how the current and voltage measurements are obtained.

variable voltage power supply

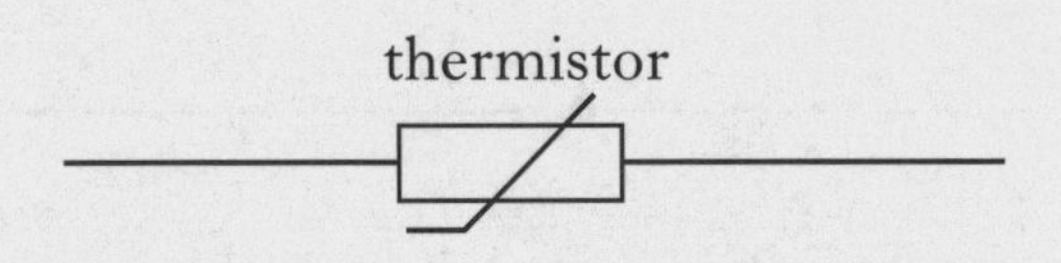

3

(*b*) The current and voltage measurements obtained are used to draw the graph shown.

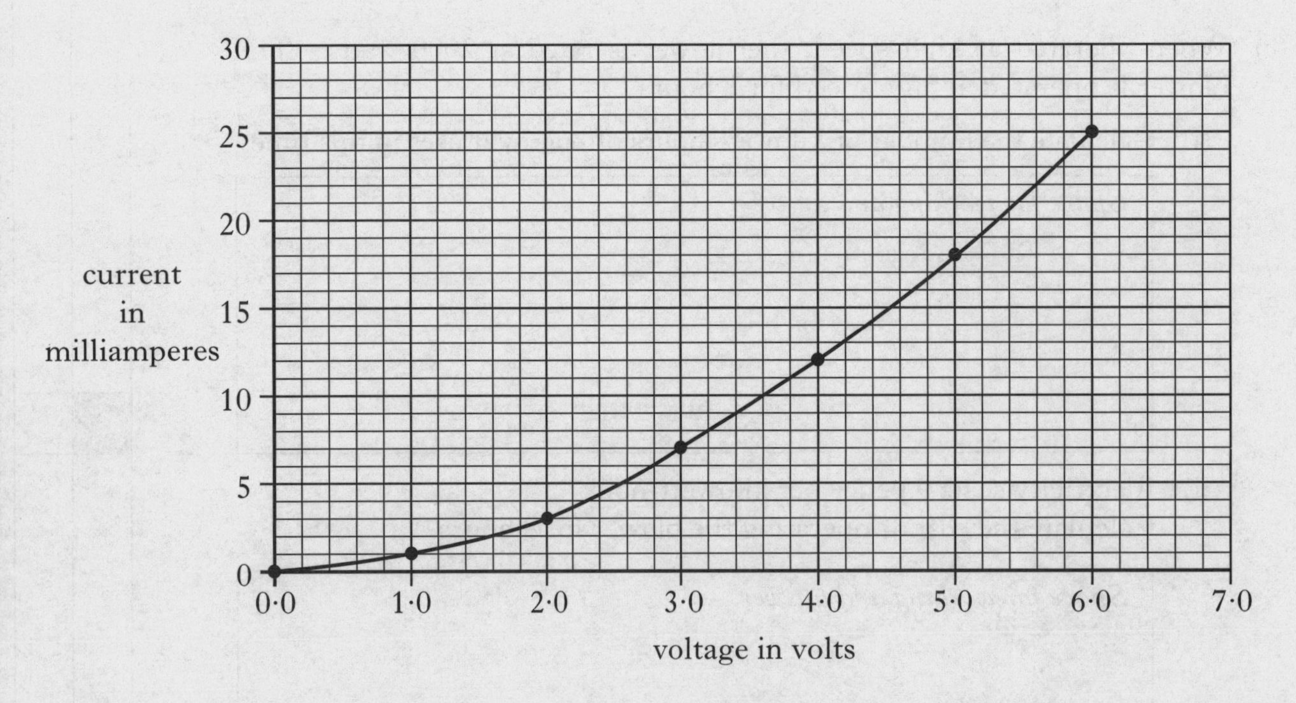

(i) What is the current in the thermistor when the voltage across the thermistor is 5·0 volts?

....

1

DO NOT WRITE IN THIS MARGIN

K&U	PS

Marks

10. (b) (continued)

(ii) Calculate the resistance of the thermistor when the voltage across the thermistor is 5·0 volts.

Space for working and answer

2

(iii) How does the resistance of the thermistor change as the voltage across the thermistor increases?

.... 1

[Turn over

DO NOT WRITE IN THIS MARGIN

K&U PS

Marks

11. A stethoscope is used to listen to sounds made inside a body. The diagram below shows the main parts of a stethoscope.

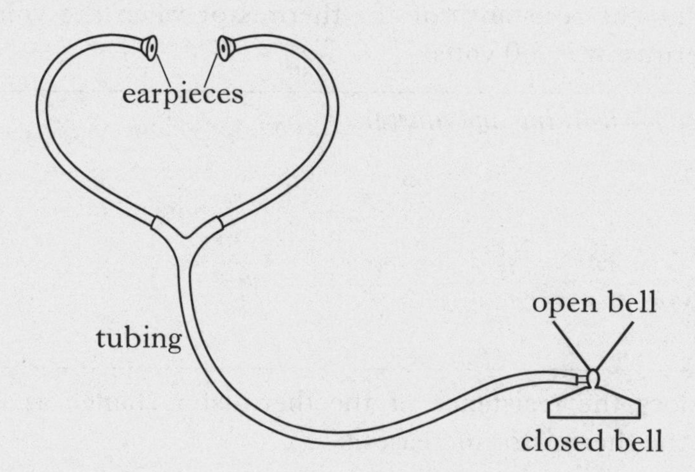

The open or closed bell is placed on the body to detect sounds.

The open bell is used for listening to heart sounds.

The graph shows how the sound level varies with the frequency of the sound detected by the bell.

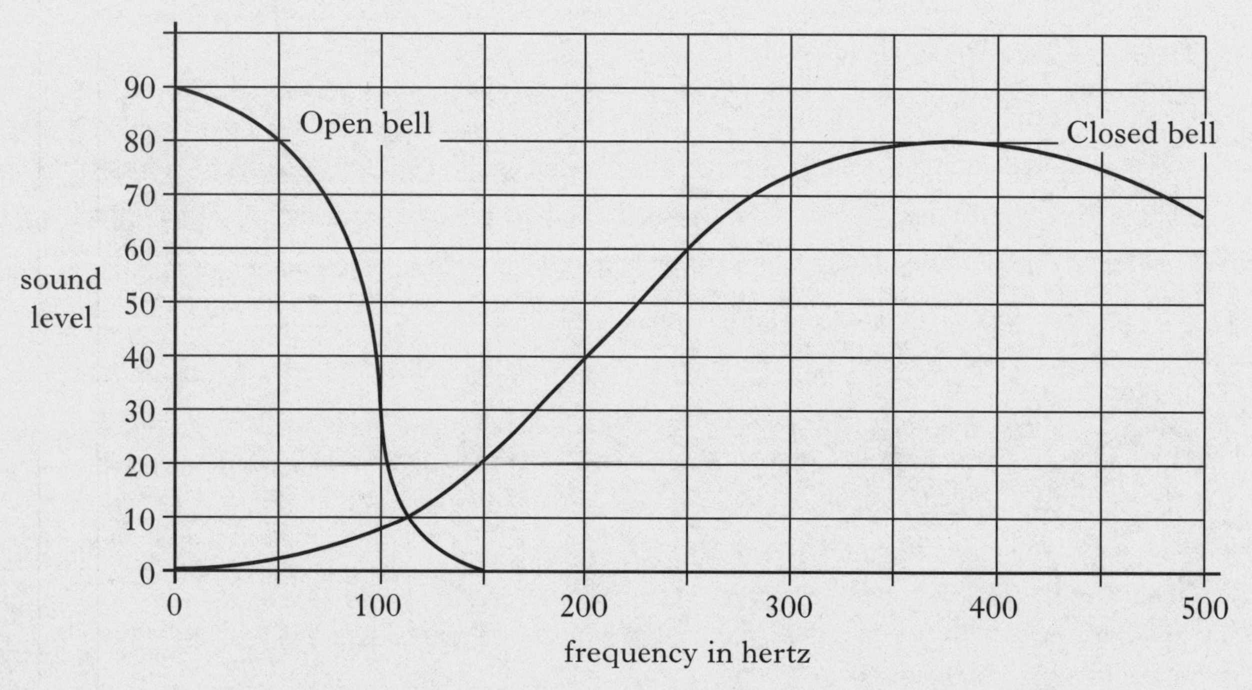

(*a*) The unit used to measure sound level has been omitted from the graph.

What is the unit of sound level?

.... **1**

DO NOT WRITE IN THIS MARGIN

K&U	PS

Marks

11. (continued)

(*b*) **Using information given**, explain whether heart sounds are high or low frequency sounds.

....

....

.... **1**

(*c*) Why is it important that the earpieces of the stethoscope fit tightly in the ears?

....

....

.... **1**

[Turn over

DO NOT WRITE IN THIS MARGIN

K&U | PS

Marks

12. The diagram shows a machine that emits nuclear radiation to treat bone cancer.

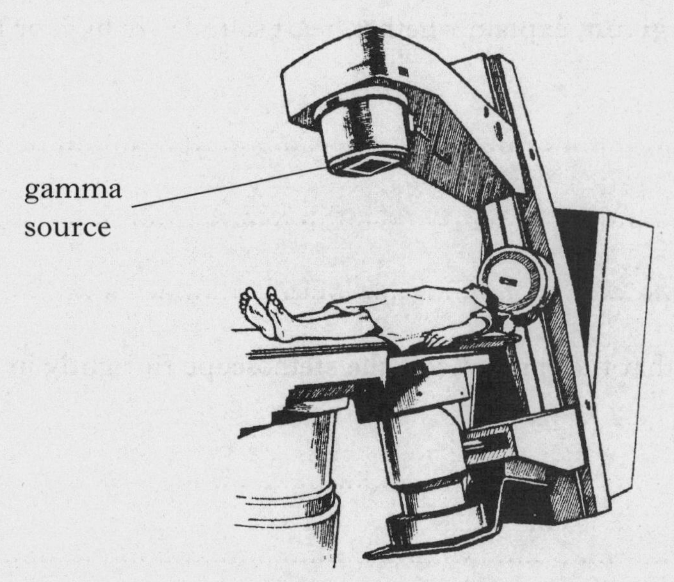

(*a*) Cancer cells are living cells that grow abnormally.

(i) What is the effect of nuclear radiation on cancer cells?

....

.... **1**

(ii) The machine emits gamma radiation.

Explain why gamma radiation is used rather than alpha or beta radiation.

....

....

.... **1**

(iii) Explain why the gamma radiation source is rotated around the patient.

....

....

.... **1**

DO NOT WRITE IN THIS MARGIN

K&U PS

Marks

12. (continued)

(*b*) The nurse who operates the machine wears a film badge containing a small piece of photographic film.

What effect does nuclear radiation have on photographic film?

....

.... **1**

[Turn over

DO NOT WRITE IN THIS MARGIN

K&U PS

Marks

13. The table below lists the upper and lower frequency limits that apply to the hearing range of different animals.

Animal	*Frequency of lower limit of hearing* (hertz)	*Frequency of upper limit of hearing* (hertz)
bat	16 000	120 000
mouse	500	60 000
cat	500	30 000
human	20	20 000

(*a*) What is the highest frequency that can be heard by a mouse?

.... 1

(*b*) Which animal **mentioned in the table** can hear the greatest range of frequencies?

.... 1

(*c*) Animals are annoyed by loud sounds within their hearing range. A householder wants to get rid of mice using an ultrasound emitter.

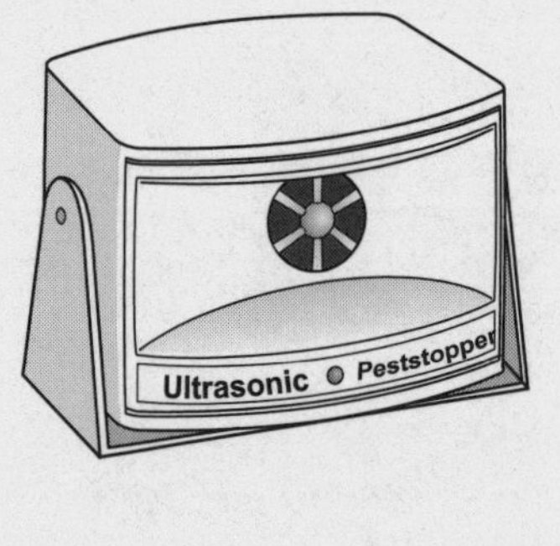

(i) What is meant by ultrasound?

....

.... 1

(ii) The householder does not want to annoy cats.

Suggest a frequency that the ultrasound emitter could operate at.

.... 1

DO NOT WRITE IN THIS MARGIN

K&U PS

Marks

14. A hearing aid is an electronic system.

(*a*) An electronic system can be represented by a block diagram as shown.

Complete the block diagram by filling in the missing label. **1**

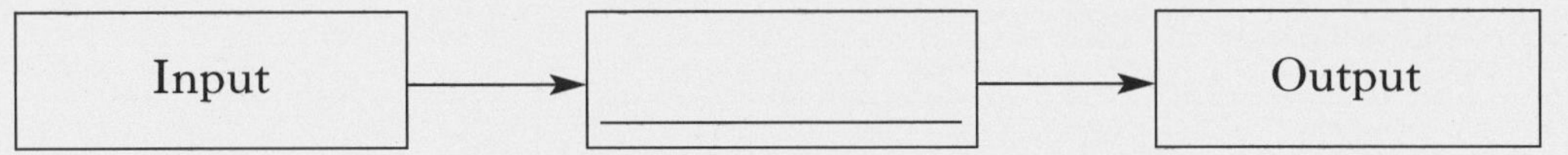

(*b*) The hearing aid contains an input device, an amplifier and an output device.

(i) Select a suitable device **from the list below** to be used as the input.

light dependent resistor **switch** **microphone**

thermocouple **solar cell** **thermistor**

.... **1**

(ii) The output device transforms electrical energy into sound energy.

State a suitable device to be used as the output.

.... **1**

(*c*) During a test of the hearing aid, the voltage generated by the input device is 0·15 volt. The voltage across the output device is 0·45 volt.

Calculate the voltage gain of the amplifier.

Space for working and answer **2**

(*d*) The voltage gain of the amplifier is now set at 5. The input signal to the amplifier is shown below.

Draw the output signal from the amplifier using the same scales. **1**

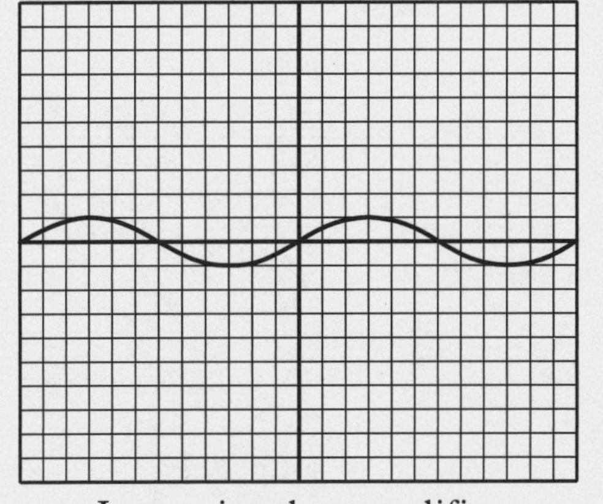

Input signal to amplifier

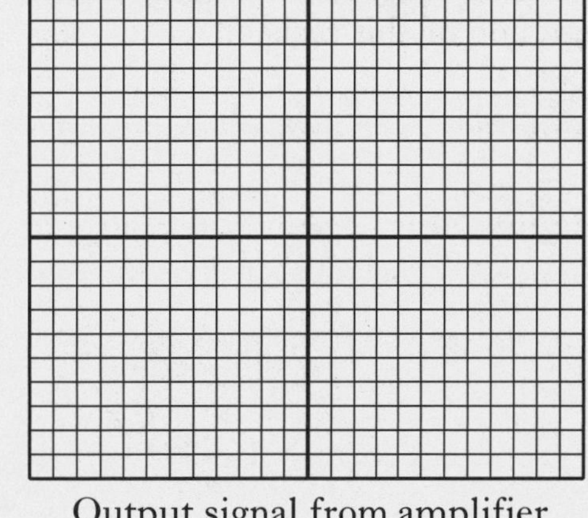

Output signal from amplifier

DO NOT WRITE IN THIS MARGIN

K&U	PS

Marks

15. A seven-segment display contains seven light emitting diodes (LEDs) arranged as shown in the diagram. Numbers are displayed by switching segments on or off.

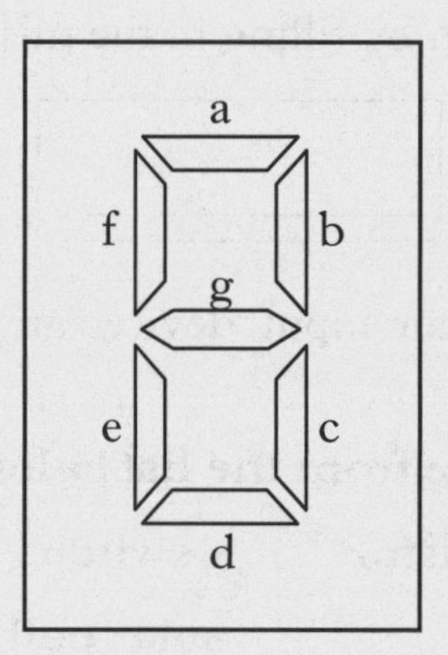

Seven-segment display

(*a*) What number is shown on the seven-segment display when only segments a, c, d, f and g are switched on?

.... **1**

(*b*) Draw the circuit symbol for an LED.

Space for circuit symbol

1

(*c*) In use, each LED is connected in series with a resistor. State the function of this resistor.

....

.... **1**

DO NOT WRITE IN THIS MARGIN

K&U PS

Marks

16. A car travels forwards along a level road at a constant speed.

(*a*) Label the diagram to show the horizontal forces acting on the car.

You must indicate the direction of each force.

2

(*b*) The car brakes suddenly.

(i) Explain, in terms of forces, why it is important for the passengers to be wearing seat belts.

....

....

.... 2

(ii) A force of 8000 newtons stops the car when the brakes are applied. The mass of the car is 1000 kilograms. The car stops in a distance of 23 metres.

(A) Calculate the acceleration of the car as it comes to rest.

Space for working and answer

2

(B) How much work is done stopping the car?

Space for working and answer

2

(iii) What is the main energy transformation in the car brakes?

....

.... 1

DO NOT WRITE IN THIS MARGIN

K&U	PS

Marks

17. Electricity can be generated from different energy sources.

(*a*) Coal is a fossil fuel that is used to generate electricity.

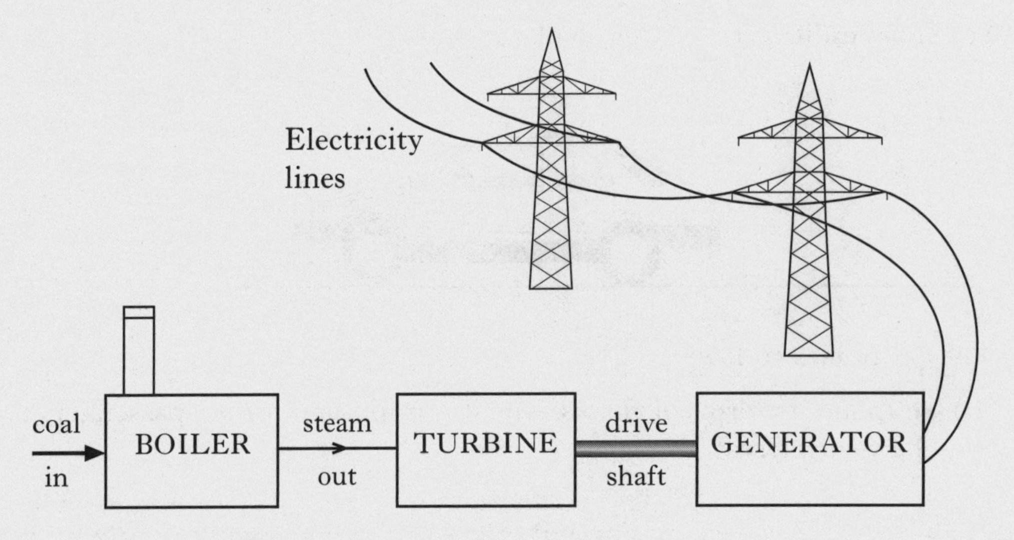

(i) In a coal-fired power station, identify the energy transformation in the boiler, the turbine and the generator.

boiler: ....to ....

turbine: ....to ....

generator: ....to **3**

(ii) State **one** disadvantage of using fossil fuels to generate electricity.

....

.... **1**

(*b*) Electricity can also be generated in nuclear power stations.

State **one** disadvantage of using nuclear fuel.

....

.... **1**

DO NOT WRITE IN THIS MARGIN

K&U	PS

Marks

18. The contents of a refrigerator are kept cool by removing heat.

This happens because a chemical called a coolant evaporates as it is pumped round pipes in the refrigerator.

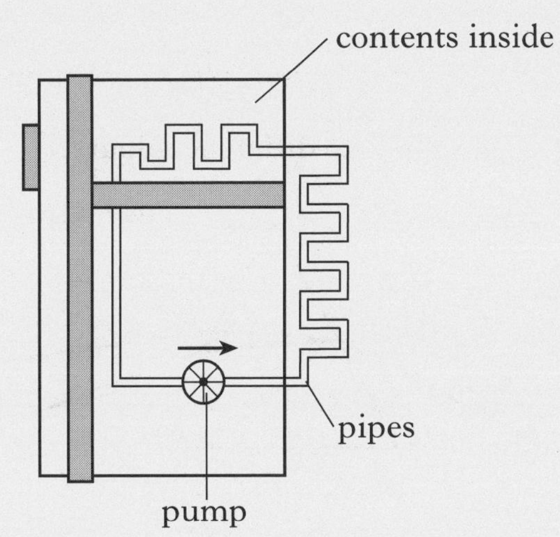

(*a*) (i) Which of the following changes of state of the coolant is used to remove heat from the contents?

A Gas to liquid

B Liquid to solid

C Liquid to gas

Answer ☐ **1**

(ii) Explain why this change of state removes heat from the contents of the refrigerator.

....

.... **1**

(*b*) A bottle containing 0·75 kilogram of milk at 22 degrees celsius is cooled in the refrigerator to 5 degrees celsius.

Calculate how much energy is removed from the milk.

[The specific heat capacity of milk is 4000 joules per kilogram per degree celsius.]

Space for working and answer

3

DO NOT WRITE IN THIS MARGIN

K&U PS

Marks

19. White light is part of the electromagnetic spectrum–a family of waves with different wavelengths.

(*a*) What property do all these waves have in common?

.... **1**

(*b*) White light can be split into different colours.

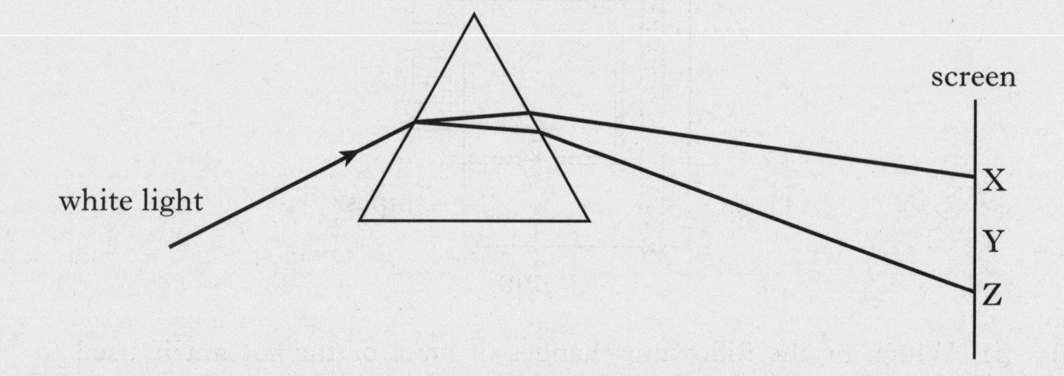

(i) What is the name of the glass block that is used to split light into different colours?

.... **1**

(ii) The colours appear on the screen in order of wavelength. The colour with the longest wavelength appears at X.

Which of the colours blue, green and red is seen on the screen at each position X, Y and Z?

X..

Y..

Z.. **2**

(iii) Which of the colours blue, green and red has the highest **frequency**?

.... **1**

DO NOT WRITE IN THIS MARGIN

K&U PS

Marks

20. The table below gives some information about planets and other objects in our Solar System.

	Distance from the Sun (million kilometres)	*Weight of 1 kilogram at the surface (newtons)*
Sun	0	270
Mercury	58	4
Venus	110	9
Earth	150	10
Moon	150	1·6
Mars	228	4
Jupiter	780	26
Saturn	1430	11
Neptune	4500	12

(*a*) Name **one** object in the table that is **not** a planet.

.... **1**

(*b*) Which planet is nearest to Earth?

.... **1**

(*c*) On which **two** planets would a 5 kilogram mass have the same weight?

....

.... **1**

[Turn over for Question 21 on *Page twenty-two*

DO NOT WRITE IN THIS MARGIN

K&U	PS

Marks

21. A spacecraft is fitted with a motor that uses electrical energy generated from sunlight. The motor is designed to propel the spacecraft from the Earth to the Moon. The mass of the spacecraft is 420 kilograms.

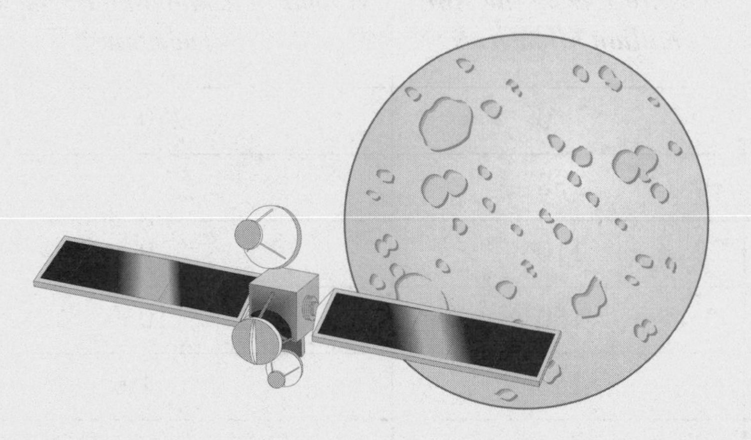

(*a*) Name a suitable device that can be used to transform light into electrical energy.

.... **1**

(*b*) The spacecraft has an acceleration of 0·2 millimetre per second per second when the motor is first switched on.

Calculate the thrust acting on the spacecraft.

Space for working and answer

2

(*c*) The motor provides thrust for the spacecraft by expelling gas at very high speed. Explain why the spacecraft moves forward when the gas is expelled.

....

....

.... **1**

[*END OF QUESTION PAPER*]

DO NOT WRITE IN THIS MARGIN

K&U	PS

YOU MAY USE THE SPACE ON THIS PAGE TO REWRITE ANY ANSWER YOU HAVE DECIDED TO CHANGE IN THE MAIN PART OF THE ANSWER BOOKLET. TAKE CARE TO WRITE IN CAREFULLY THE APPROPRIATE QUESTION NUMBER.

DO NOT WRITE IN THIS MARGIN

K&U	PS

YOU MAY USE THE SPACE ON THIS PAGE TO REWRITE ANY ANSWER YOU HAVE DECIDED TO CHANGE IN THE MAIN PART OF THE ANSWER BOOKLET. TAKE CARE TO WRITE IN CAREFULLY THE APPROPRIATE QUESTION NUMBER.

[BLANK PAGE]

FOR OFFICIAL USE

G

K & U	PS

Total Marks

3220/401

NATIONAL QUALIFICATIONS 2006

WEDNESDAY, 17 MAY
9.00 AM – 10.30 AM

PHYSICS
STANDARD GRADE
General Level

Fill in these boxes and read what is printed below.

Full name of centre

Town

Forename(s)

Surname

Date of birth
Day Month Year

Scottish candidate number

Number of seat

Reference may be made to the Physics Data Booklet.

1 All questions should be answered.

2 The questions may be answered in any order but all answers must be written clearly and legibly in this book.

3 For questions 1–7, write down, in the space provided, the letter corresponding to the answer you think is correct. There is only **one** correct answer.

4 For questions 8–19, write your answer where indicated by the question or in the space provided after the question.

5 If you change your mind about your answer you may score it out and replace it in the space provided at the end of the answer book.

6 Before leaving the examination room you must give this book to the invigilator. If you do not, you may lose all the marks for this paper.

SA 3220/401 6/22770

DO NOT WRITE IN THIS MARGIN

K&U	PS

Marks

1. Which part of a television receiver picks up all signals?

A Tuner

B Modulator

C Decoder

D Amplifier

E Aerial

Answer ☐ **1**

2. The nucleus of a uranium atom contains

A electrons only

B neutrons only

C electrons and protons only

D protons and neutrons only

E electrons, protons and neutrons.

Answer ☐ **1**

3. What is the unit of equivalent dose?

A becquerel

B joule

C kilogram

D sievert

E watt

Answer ☐ **1**

DO NOT WRITE IN THIS MARGIN

K&U	PS

Marks

4. An uncharged capacitor C is connected to a resistor R, a 9 volt battery and a switch S as shown.

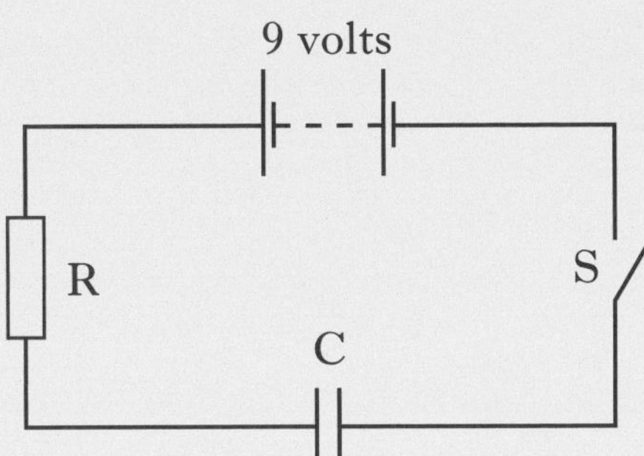

When switch S is closed the voltage across the capacitor

A remains at 0 volt

B gradually rises from 0 volt to 9 volts

C immediately drops from 9 volts to 0 volt

D gradually drops from 9 volts to 0 volt

E remains at 9 volts.

Answer ☐ **1**

5. Which of the following is a unit of heat?

A degree celsius

B joule

C joule per kilogram

D joule per kilogram per degree celsius

E watt

Answer ☐ **1**

6. Which of the following is the shortest distance?

The distance from the Earth to the

A nearest star in our galaxy

B edge of our galaxy

C Moon

D Sun

E nearest planet.

Answer ☐ **1**

DO NOT WRITE IN THIS MARGIN

K&U	PS

Marks

7. Radio waves from space can be detected by a

A Geiger-Müller tube

B photographic plate

C scintillation counter

D telescope

E tuner.

Answer ☐ **1**

DO NOT WRITE IN THIS MARGIN

K&U	PS

Marks

8. A factory chimney is demolished using explosives.

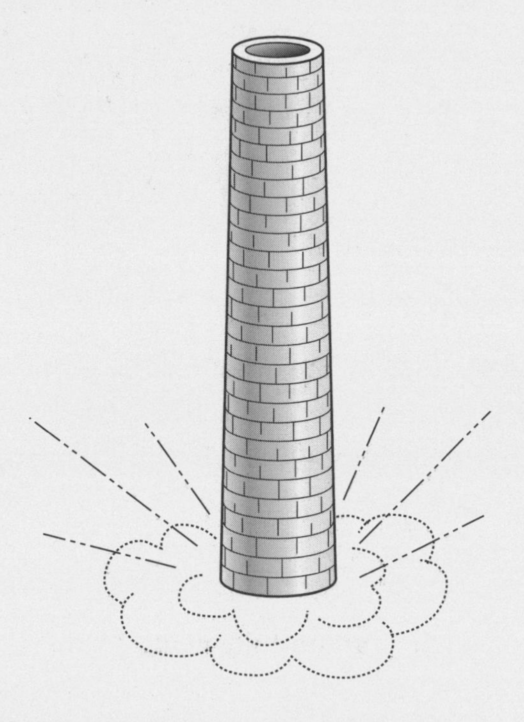

A crowd of people watches from a safe distance. A person in the crowd hears the sound 2·5 seconds after seeing the explosion.

(*a*) Explain why there is a delay between seeing the explosion and hearing the sound.

....

.... **1**

(*b*) Calculate the distance between the chimney and the person in the crowd. (The speed of sound in air is 340 metres per second.)

Space for working and answer

2

(*c*) Why should the demolition worker who sets off the explosives wear ear protectors to reduce the noise level to below 80 decibels?

....

....

.... **2**

DO NOT WRITE IN THIS MARGIN

K&U PS

Marks

9. The flex of a mains appliance has a 3-pin plug fitted as shown.

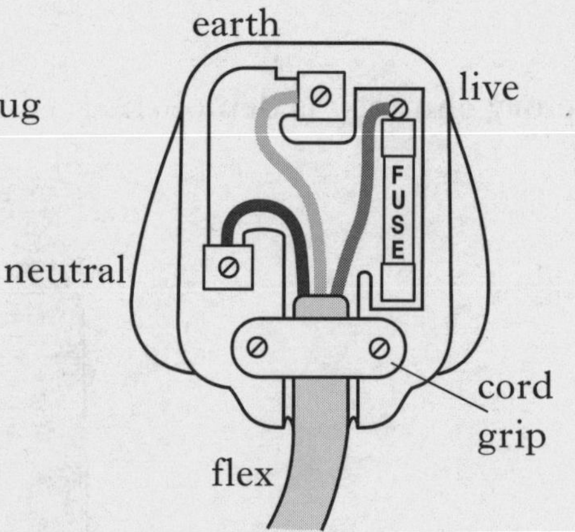

The flex contains three wires—live, neutral and earth.

(*a*) Circle the correct answer for each of the questions about the wires.

(i) The colour of the insulation around the live wire is { blue / brown / green/yellow }. **1**

(ii) The colour of the insulation around the neutral wire is { blue / brown / green/yellow }. **1**

(iii) The { earth / live / neutral } wire is a safety device. **1**

(*b*) **Explain** why the flex must be held in place by the cord grip.

....

....

.... **2**

(*c*) Another appliance has only two wires in its flex. This appliance carries the following symbol.

(i) Name this symbol.

.... **1**

(ii) Which wire is not needed in this flex?

.... **1**

DO NOT WRITE IN THIS MARGIN

K&U	PS

Marks

10. Read the following passage.

The temperature of the human body is maintained at about 37 degrees celsius. An increase or a decrease in body temperature of as little as 5 degrees celsius can be very serious.

Doctors often use ear thermometers to measure body temperature. Ear thermometers measure the infrared radiation emitted from the eardrum and surrounding tissue.

One type of ear thermometer has a scale that ranges from 32 degrees celsius to 42 degrees celsius. The temperature sensor used in this thermometer is a device that has a resistance which changes as the temperature changes.

Use information **given in the passage** to answer the following questions.

(*a*) Name the type of radiation given out by the human body.

.... **1**

(*b*) Give a reason why the scale of the ear thermometer ranges from 32 degrees celsius to 42 degrees celsius only.

....

.... **1**

(*c*) Suggest a temperature sensor that could be used in the ear thermometer.

.... **1**

[Turn over

DO NOT WRITE IN THIS MARGIN

K&U | PS

Marks

11. A student has a sight defect and is unable to see near objects clearly.

(*a*) The following diagram shows what happens to light rays when the student is reading a book.

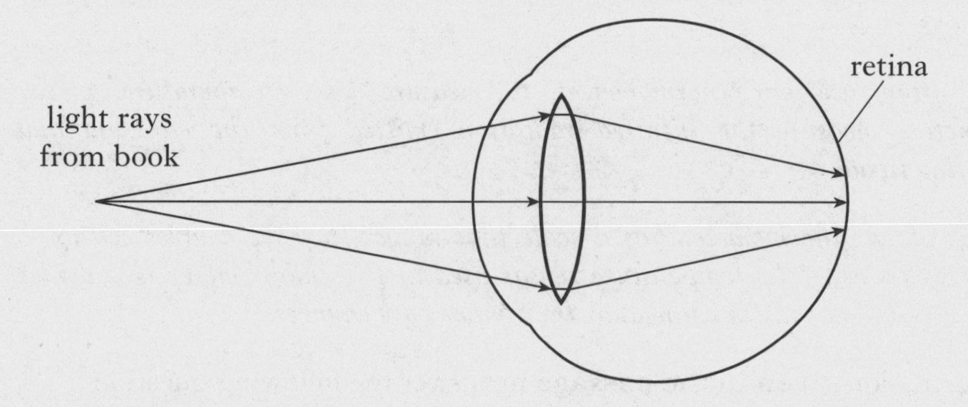

(i) By referring to the diagram, explain why the student sees a blurred image.

....

.... **1**

(ii) Name this sight defect.

.... **1**

(iii) In the space below, draw the shape of the lens that would correct this sight defect.

Space for diagram

1

(iv) When this sight defect has been corrected, the student looks at a picture printed in the book.

Which statement describes the image on the retina of the student's eye compared to the actual picture?

A The image is the same way up and larger.

B The image is upside down and larger.

C The image is the same way up and smaller.

D The image is upside down and smaller.

Answer ☐ **1**

DO NOT WRITE IN THIS MARGIN

K&U	PS

Marks

11. (continued)

(*b*) Another student has a different eye defect. This student is prescribed spectacles that have red tinted glass. The graph below shows the percentage of light of different colours that passes through this glass.

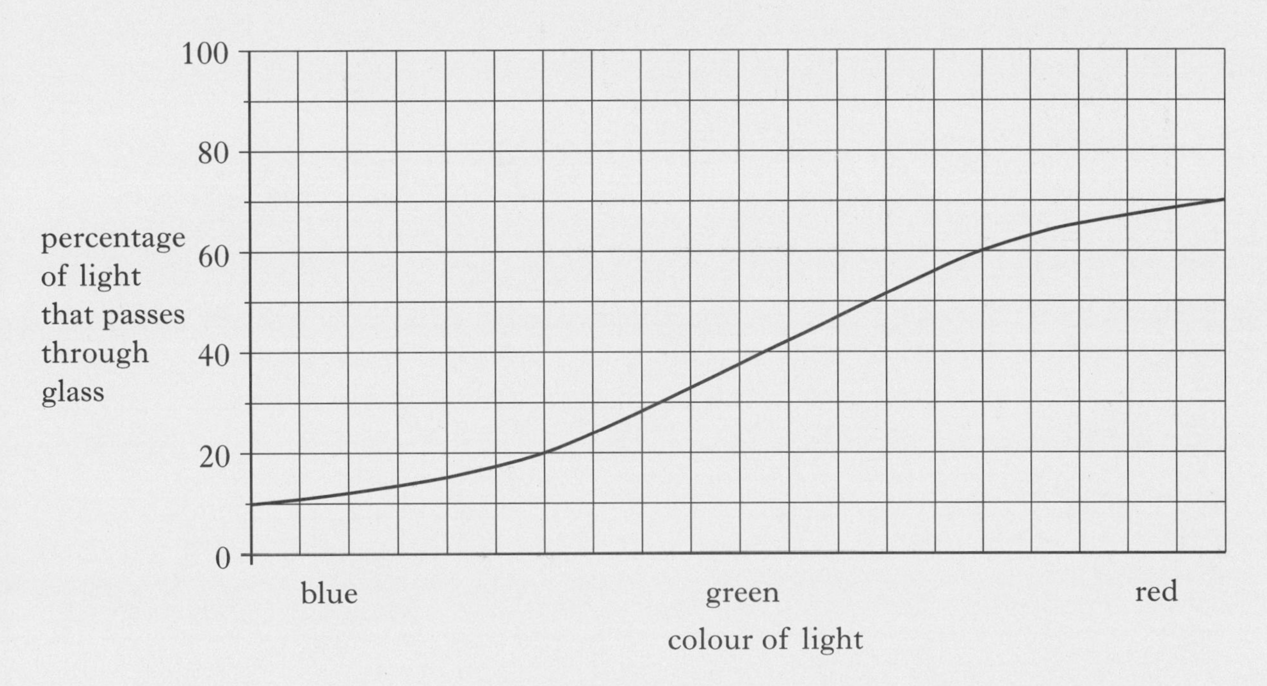

(i) Which colour of light is blocked most by the tinted glass?

.... **1**

(ii) List the three colours given on the graph in order of **decreasing** wavelength.

.... **1**

[Turn over

DO NOT WRITE IN THIS MARGIN

K&U	PS

Marks

12. A karaoke machine contains various input and output devices.

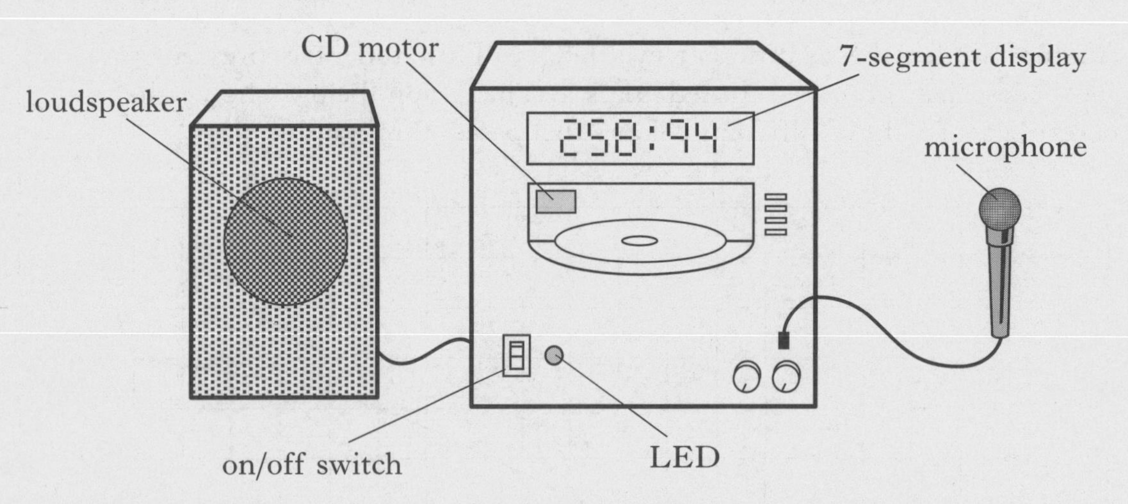

(*a*) State **two** output devices labelled on the diagram.

Device 1 ....

Device 2 .... **2**

(*b*) State **two** input devices labelled on the diagram.

Device 1 ....

Device 2 .... **2**

(*c*) The karaoke machine has an LED.

(i) State the useful energy transfer that takes place in the LED.

.... to **1**

(ii) In the space below draw the symbol for an LED.

Space for symbol

1

DO NOT WRITE IN THIS MARGIN

K&U PS

Marks

13. A technician uses a signal generator and two oscilloscopes as shown to test an amplifier.

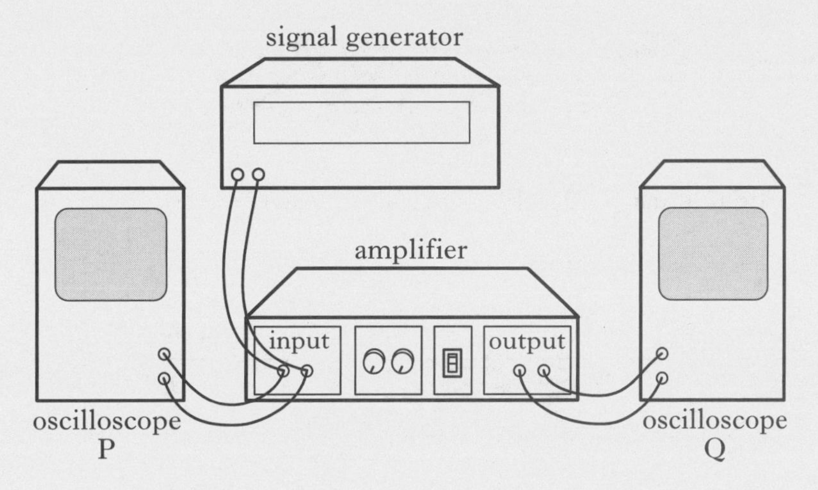

(*a*) The screens of both oscilloscopes are shown below.

input signal to amplifier

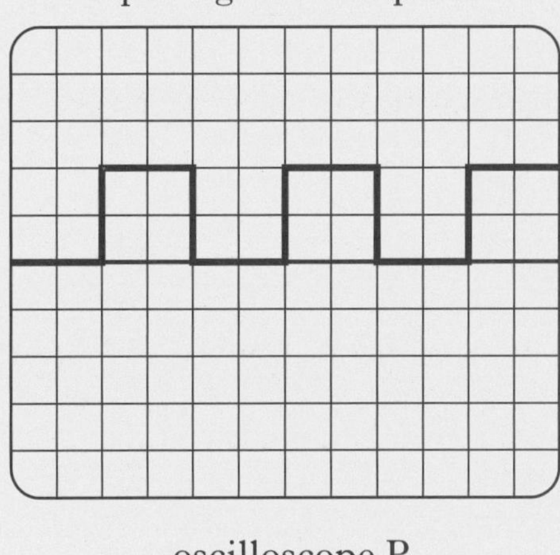

oscilloscope P

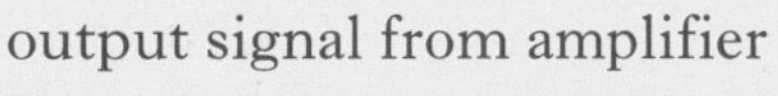
output signal from amplifier

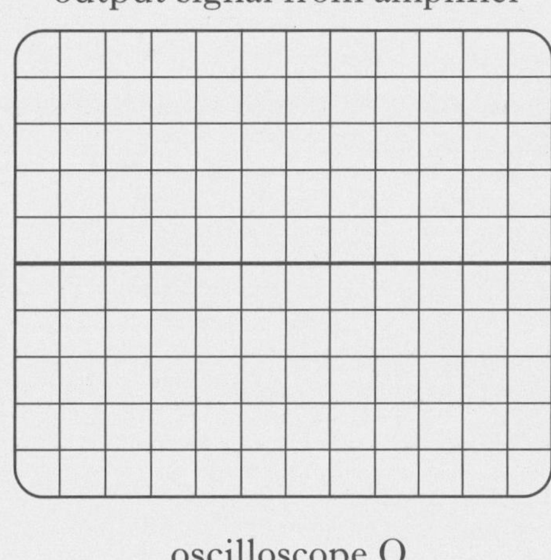

oscilloscope Q

The settings on both oscilloscopes are identical.

(i) Complete the diagram to show the amplified output signal seen on oscilloscope Q. **2**

(ii) Circle the correct answer in the statement below.

The signal shown on oscilloscope P is { analogue / decimal / digital } . **1**

(*b*) Which of the following devices contains an amplifier?

lamp **radio** **relay** **transformer**

.... **1**

DO NOT WRITE IN THIS MARGIN | K&U | PS

Marks

14. A rowing crew takes part in a race.

The time for their boat at each stage of the race is shown.

		Time from start	
		minutes	*seconds*
Start:	0 metres	00	00
	500 metres	01	40
	1000 metres	03	50
	1500 metres	05	50
Finish:	2000 metres	07	45

(*a*) **Describe** how to find the average speed of the boat from the start of the race to the finish.

....

....

....

....

.... **3**

(*b*) Calculate the average speed of the boat during the first 500 metres of the race.

Space for working and answer

2

DO NOT WRITE IN THIS MARGIN

K&U PS

Marks

14. (continued)

(*c*) The crew supplies a force to move the boat forward. When the boat is moving, a force opposes the motion of the boat.

(i) Name the force that opposes the motion of the boat.

.... 1

(ii) During the first 500 metres, there is a constant unbalanced force acting on the boat.

Describe the motion of the boat during this section of the race.

.... 1

(iii) During one stage of the race, the speed of the boat is constant.

What can be said about the forces acting on the boat during this stage?

.... 1

[Turn over

DO NOT WRITE IN THIS MARGIN

K&U	PS

Marks

15. A car is being repaired in a garage. The car is on a ramp and is raised to a height of 1·5 metres.

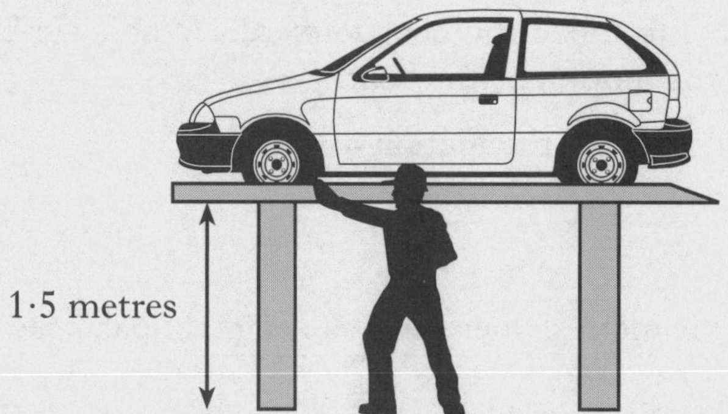

The car has a mass of 1200 kilograms.

(*a*) Calculate the weight of the car.

Space for working and answer

2

(*b*) Calculate how much gravitational potential energy the car has gained when it is 1·5 metres above the garage floor.

Space for working and answer

2

(*c*) The car is raised in 12 seconds.

(i) Calculate the minimum power needed to lift the car 1·5 metres in 12 seconds.

Space for working and answer

2

(ii) In practice, the power needed to raise the car in this time is greater than the minimum power.

Explain why.

....

.... **1**

DO NOT WRITE IN THIS MARGIN

K&U	PS

Marks

16. A fan operates using a solar cell and a light bulb.

60 watts

mains

solar cell

fan

(*a*) What energy transformation takes place in the **solar cell**?

.... to **1**

(*b*) When the lamp is on, the fan turns slowly.

(i) Suggest **two** changes that could be made which would make the fan turn faster.

Change 1 ....

Change 2 .... **2**

(ii) The 60 watt lamp operates for 2 minutes.

Calculate how much energy is transformed by the lamp in this time.

Space for working and answer

2

(*c*) Solar energy is a renewable source of energy.

(i) Name **one** other renewable source of energy.

.... **1**

(ii) Name a non-renewable source of energy.

.... **1**

[Turn over

DO NOT WRITE IN THIS MARGIN

K&U PS

Marks

17. The diagram shows all the ways in which heat is lost from a house.

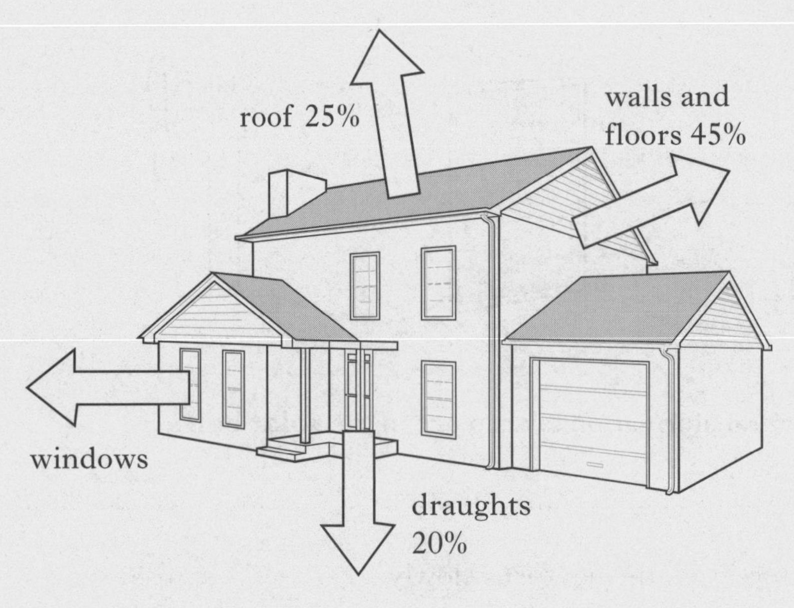

(*a*) Using information from the diagram, calculate the percentage of heat lost through windows.

Space for working and answer

2

(*b*) Various windows of area one square metre are tested for rate of heat loss. The results are shown in the table.

Window	*Rate of heat loss* (joules per second)
single glazed	80
double glazed	60
triple glazed	50

(i) How many joules of heat are lost per square metre from a single glazed window every second?

.... 1

DO NOT WRITE IN THIS MARGIN

K&U | PS

Marks

17. (*b*) (continued)

(ii) All the windows in a particular house are single glazed. Every second a total of 500 joules of heat is lost through the windows in this house.

(A) Calculate the total area of the windows.

Space for working and answer

2

(B) Describe **one** way of reducing heat loss through the windows in this house.

....

.... 1

(*c*) A householder keeps the temperature in a house at 20 degrees celsius all year.

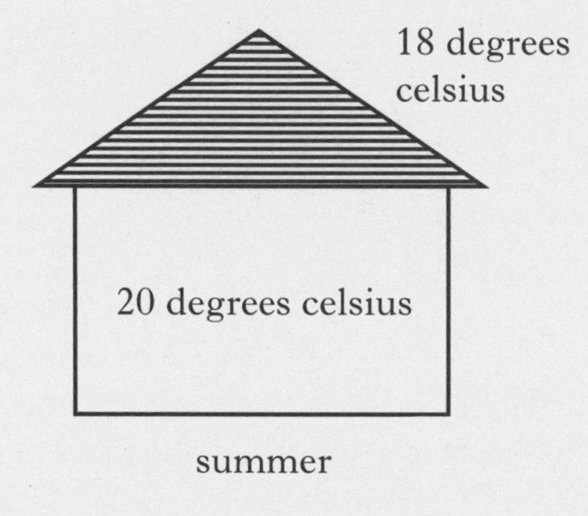

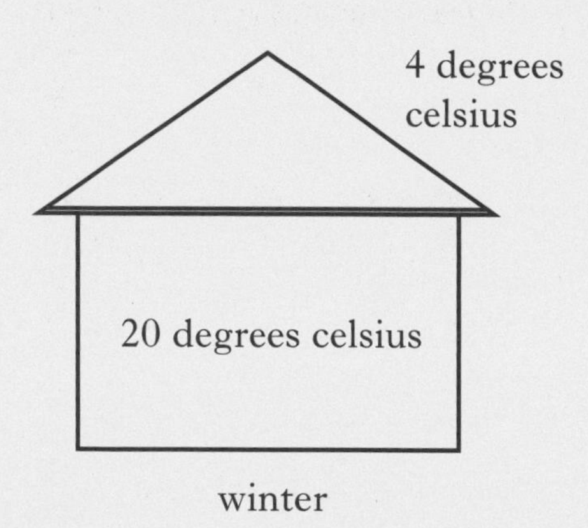

At which time of the year is the rate of heat loss from this house greater?

Explain your answer.

....

....

.... 2

DO NOT WRITE IN THIS MARGIN

K&U | PS

Marks

18. A 5 volt battery in a mobile phone is recharged from the mains using a charger containing a step down transformer.

(*a*) The transformer consists of three parts.

core **primary coil** **secondary coil**

Label each of these parts on the diagram below.

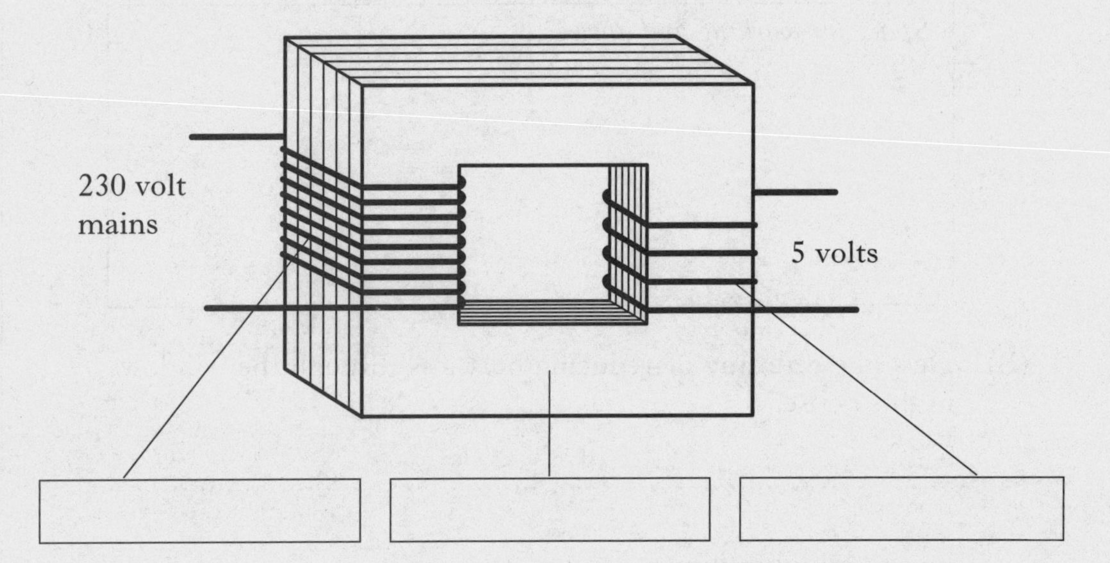

3

(*b*) There are 11 500 turns on the primary coil of the transformer.

Calculate the number of turns on the secondary coil.

Space for working and answer

2

(*c*) **Explain** why a transformer cannot be used to step down the voltage from a battery.

....

....

.... 2

Marks

19. A spacecraft is far out in space. An astronaut wearing a backpack leaves the spacecraft. The astronaut uses the backpack to move around. The backpack contains a pressurised gas cylinder connected to a valve. When the valve is opened, a jet of gas is released.

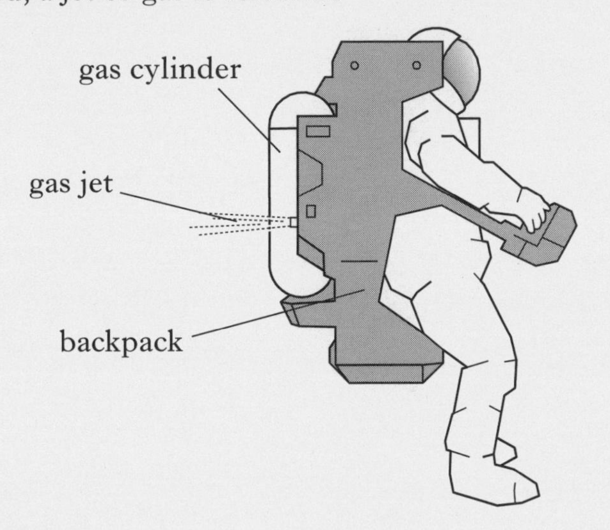

(*a*) Complete the passage below by circling the correct answer.

When the astronaut opens the valve, the cylinder pushes gas backwards.

The gas pushes the { cylinder / jet / spacecraft } forwards. **1**

(*b*) The astronaut and backpack have a combined mass of 120 kilograms. The jet of gas exerts a constant thrust of 24 newtons.

(i) Calculate the acceleration of the astronaut when the jet is switched on.

Space for working and answer

2

(ii) The jet is now switched off.

Describe the motion of the astronaut.

Explain your answer.

....

....

.... **2**

[*END OF QUESTION PAPER*]

DO NOT WRITE IN THIS MARGIN

K&U	PS

YOU MAY USE THE SPACE ON THIS PAGE TO REWRITE ANY ANSWER YOU HAVE DECIDED TO CHANGE IN THE MAIN PART OF THE ANSWER BOOKLET. TAKE CARE TO WRITE IN CAREFULLY THE APPROPRIATE QUESTION NUMBER.

2007 | General

[BLANK PAGE]

FOR OFFICIAL USE

G

K & U	PS

Total Marks

3220/401

NATIONAL QUALIFICATIONS 2007

WEDNESDAY, 16 MAY
9.00 AM – 10.30 AM

PHYSICS
STANDARD GRADE
General Level

Fill in these boxes and read what is printed below.

Full name of centre

Town

Forename(s)

Surname

Date of birth
Day Month Year

Scottish candidate number

Number of seat

Reference may be made to the Physics Data Booklet.

1 All questions should be answered.

2 The questions may be answered in any order but all answers must be written clearly and legibly in this book.

3 For questions 1–5, write down, in the space provided, the letter corresponding to the answer you think is correct. There is only **one** correct answer.

4 For questions 6–18, write your answer where indicated by the question or in the space provided after the question.

5 If you change your mind about your answer you may score it out and replace it in the space provided at the end of the answer book.

6 Before leaving the examination room you must give this book to the invigilator. If you do not, you may lose all the marks for this paper.

SA 3220/401 6/22870

DO NOT WRITE IN THIS MARGIN

K&U	PS

Marks

1. Which part of a radio receiver separates the audio signal from the carrier wave?

A Aerial

B Tuner

C Decoder

D Amplifier

E Loudspeaker

Answer ☐ 1

2. Four **identical** resistors, P, Q, R and S are connected as shown.

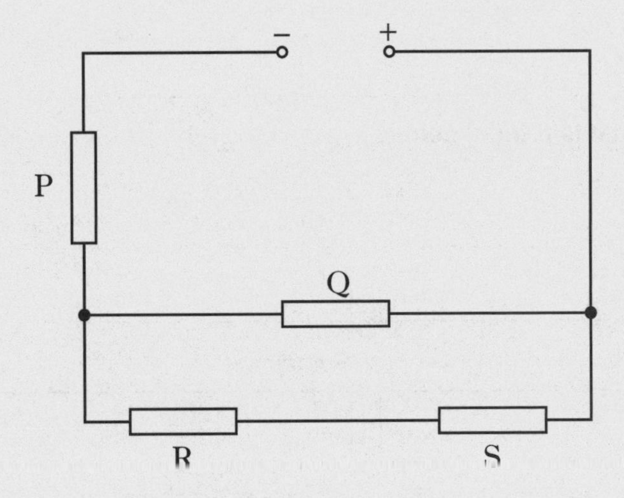

In which of the resistors is the current the same?

A P and Q only

B R and S only

C P, R and S only

D Q, R and S only

E P, Q, R and S.

Answer ☐ 1

DO NOT WRITE IN THIS MARGIN

K&U PS

Marks

3. Which row of values would result in the greatest kinetic energy?

	Mass (kilograms)	*Speed* (metres per second)
A	45	8
B	45	4
C	50	10
D	50	8
E	50	4

Answer ☐ **1**

4. A rocket is pushed forwards because its engine gases

A are pushed backwards

B spread outwards

C are pushed forwards

D surround the rocket

E spread inwards.

Answer ☐ **1**

5. In outer space, the engine of a space probe is switched on for a short time. When the engine is switched off, the rocket

A changes direction

B moves at a steady speed

C slows down

D speeds up

E follows a curved path.

Answer ☐ **1**

[Turn over

DO NOT WRITE IN THIS MARGIN

K&U PS

Marks

6. A surfer rides the waves near a beach.

(*a*) The diagram below shows a wave some distance from the beach.

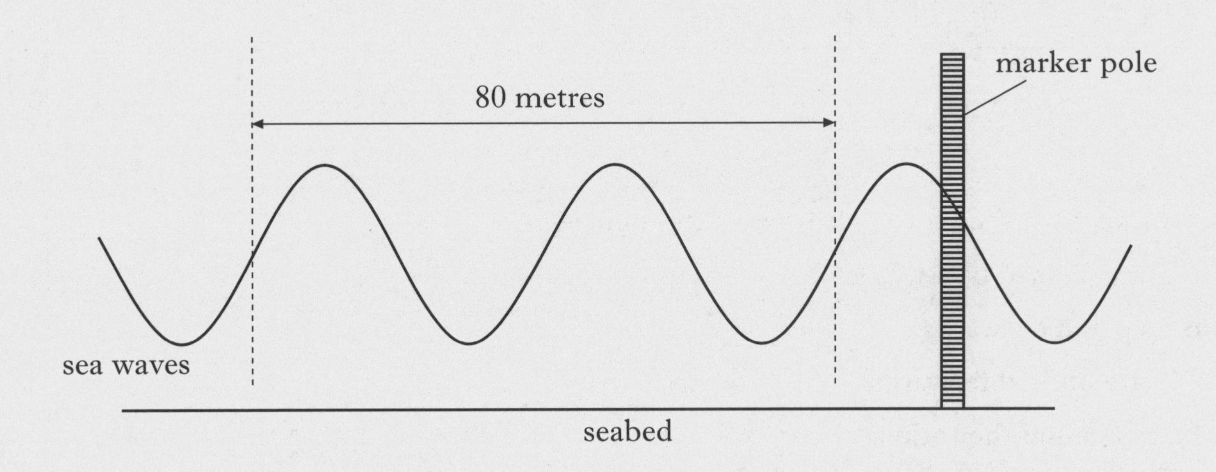

(i) Using information from the diagram, calculate the wavelength of the wave.

Space for working and answer

2

(ii) The time between one crest and the next crest passing the marker pole is 5 seconds.

Calculate the speed of the wave.

Space for working and answer

2

DO NOT WRITE IN THIS MARGIN

K&U PS

Marks

6. (*a*) (continued)

(iii) Calculate the frequency of the wave.

Space for working and answer

2

(*b*) The drawing below shows changes in the wave as it approaches the beach.

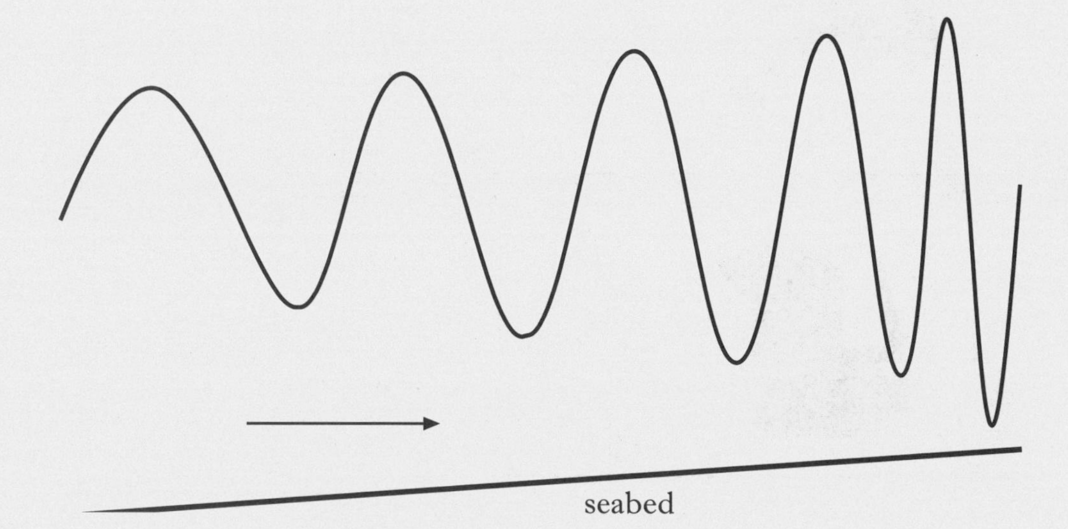

Complete the sentences below by circling the correct answers.

(i) As the wave approaches the beach,

its wavelength {decreases / increases / stays the same}. **1**

(ii) As the wave approaches the beach,

its amplitude {decreases / increases / stays the same}. **1**

[Turn over

DO NOT WRITE IN THIS MARGIN

K&U | PS

Marks

7. Appliances convert electrical energy into other forms of energy.

Appliance	*Rating plate*
Food processor	230 volts 50 hertz 400 watts
Hair dryer	230 volts 50 hertz ▣ 1200 watts
Kettle	230 volts 50 hertz 2200 watts
Lamp	230 volts 50 hertz ▣ 60 watts

(*a*) State the **useful** energy output from the following appliances.

(i) Lamp: electrical energy ⟶ .... energy **1**

(ii) Kettle: electrical energy ⟶ .... energy **1**

DO NOT WRITE IN THIS MARGIN

K&U	PS

Marks

7. (continued)

(*b*) (i) Name **one** appliance from the table which requires an earth wire.

.... 1

(ii) Circle **one** word or phrase in the passage below to make the statement correct.

The colouring of the insulation around the earth wire is

{ blue / brown / green and yellow }. 1

(iii) Each appliance is fitted with either a 3 ampere or 13 ampere fuse. State the correct value of fuse for the following appliances.

(A) Lamp: 1

(B) Hair dryer: 1

[Turn over

DO NOT WRITE IN THIS MARGIN

K&U	PS

Marks

8. A mobile phone contains a battery which is charged using a base unit. The base unit contains a transformer and is connected to the a.c. mains supply.

mobile phone containing battery

base unit containing transformer

to a.c. mains supply

(*a*) What is the purpose of the mains supply?

.... **1**

(*b*) Name the supply mentioned which is d.c.

.... **1**

(*c*) a.c. is short for alternating current.
Explain what is meant by alternating current.

....

.... **1**

(*d*) State the purpose of a transformer.

....

.... **1**

(*e*) State **one** advantage of using a mobile phone.

....

.... **1**

DO NOT WRITE IN THIS MARGIN

K&U	PS

Marks

9. One of the spotlights on a stage does not work. A continuity tester is used to find the fault. The continuity tester contains a lamp and a 1·5 volt battery.

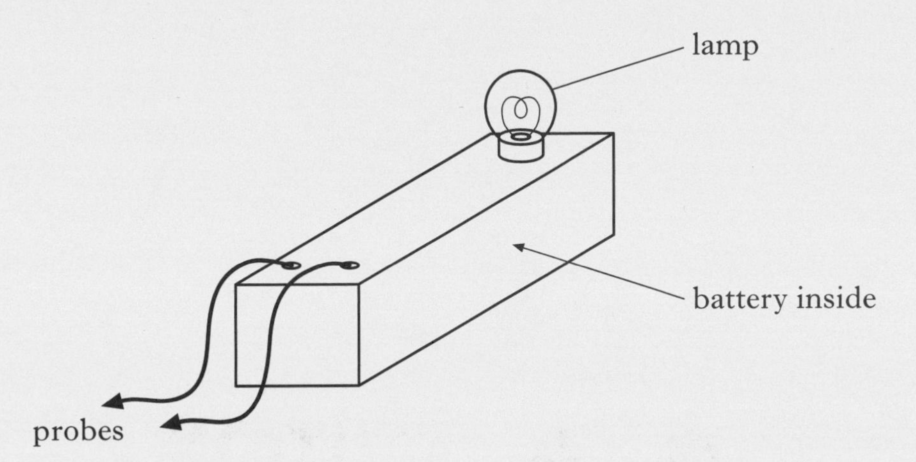

(*a*) Complete the circuit diagram for the continuity tester.

You must use the correct symbols for all components.

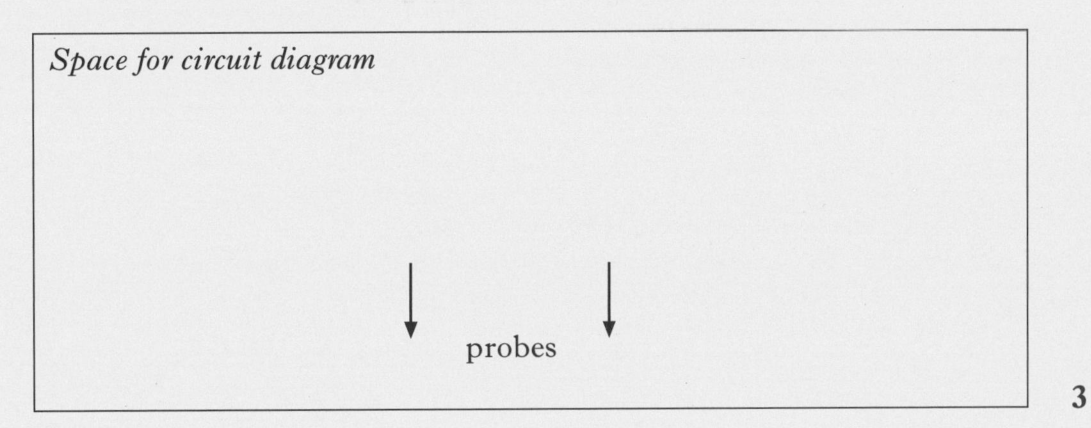

3

(*b*) Describe how you could check that the continuity tester is working properly.

....

.... 2

(*c*) The continuity tester shows that the fault in the spotlight is an open circuit.

What is meant by an open circuit?

.... 1

[Turn over

DO NOT WRITE IN THIS MARGIN

K&U	PS

Marks

10. Different types of radiation are used to detect and treat illnesses and injuries. Four of these radiations are

infrared	**laser light**	**ultraviolet**	**X-rays**

(*a*) What type of radiation is used to treat skin conditions such as acne?

.... **1**

(*b*)

(i) State **one** medical use of X-rays.

.... **1**

(ii) What can be used to detect X-rays?

.... **1**

DO NOT WRITE IN THIS MARGIN

K&U PS

Marks

10. (continued)

(*c*)

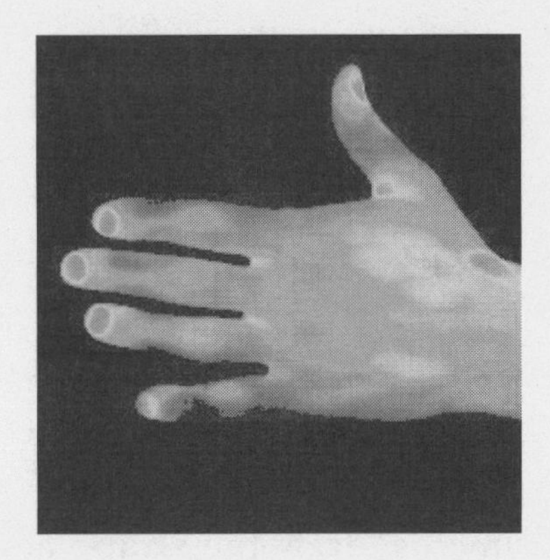

Colour photographs called thermograms are used to find the temperature variation in a patient's body.

Name the radiation used to make thermograms.

.... **1**

(*d*)

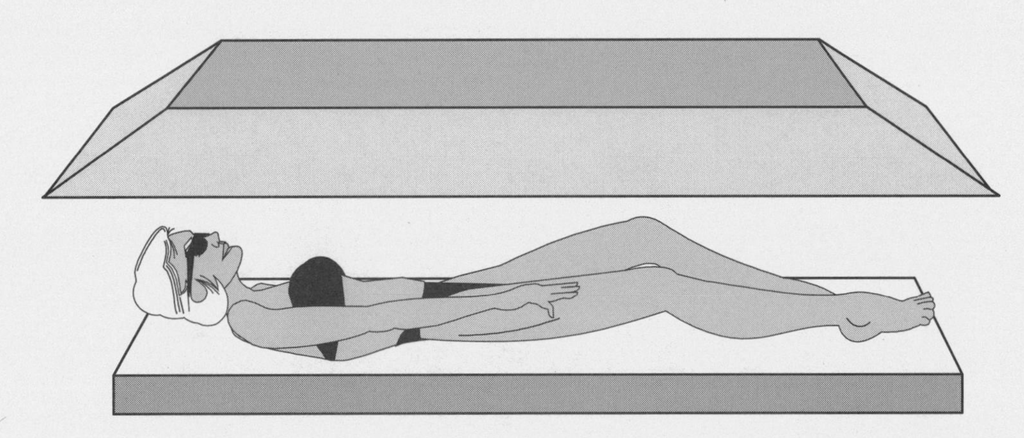

Explain why people need to be protected from overexposure to ultraviolet radiation.

.... **1**

[Turn over

DO NOT WRITE IN THIS MARGIN

K&U	PS

Marks

11. A class investigates the effects of the following shapes of glass on rays of white light.

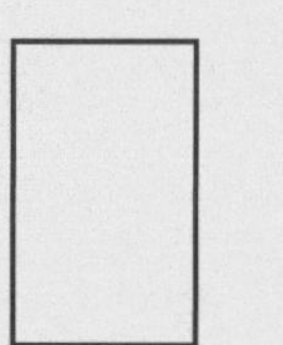

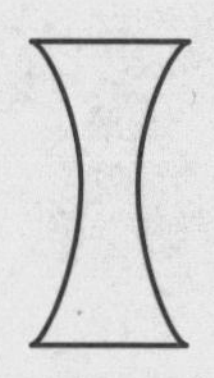

The teacher sets up three experiments, covering the glass shape with card. The paths of the light rays entering and leaving the different shapes of glass are shown.

For each of the three experiments, draw the **shape** and **position** of the glass block that was used.

(*a*)

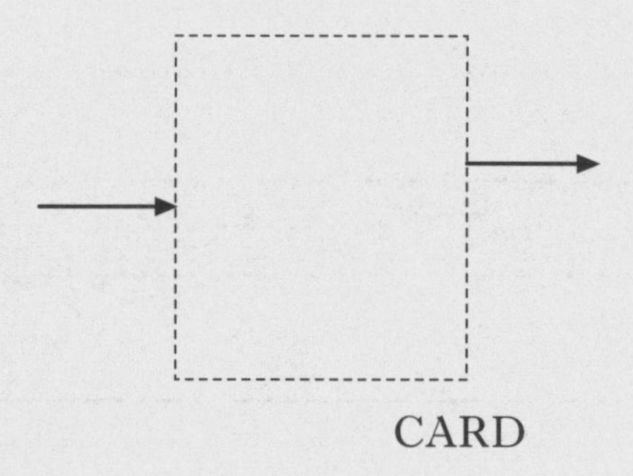

2

(*b*)

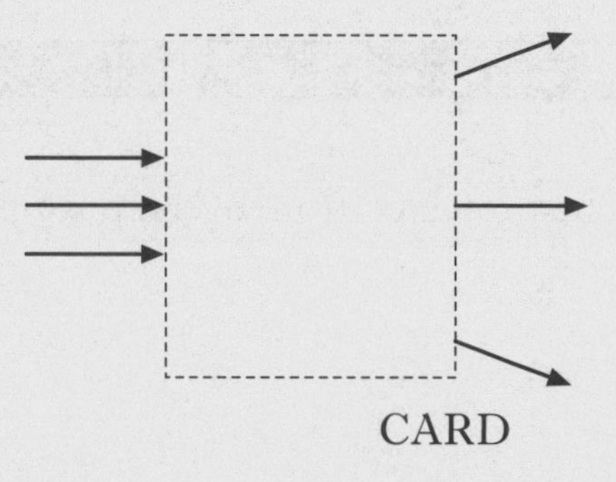

2

(*c*)

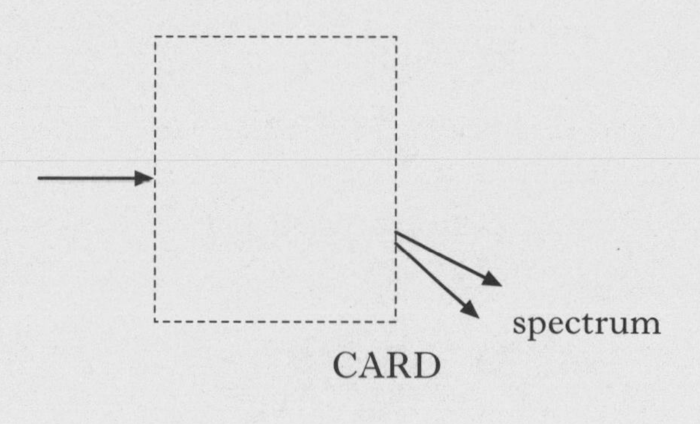

2

DO NOT WRITE IN THIS MARGIN

K&U | PS

Marks

12. A radio and a computer mouse are examples of electronic systems.

(*a*) An electronic system can be represented by a block diagram as shown.

Complete the block diagram by filling in the missing labels.

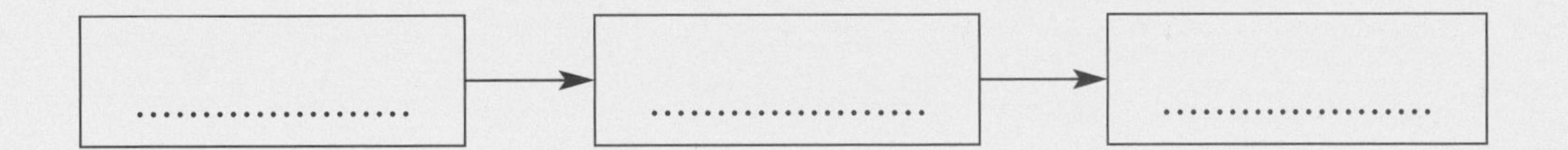

3

(*b*) Output signals from an electronic system can be either analogue or digital.

(i) The output signal from a radio is analogue.

Draw an analogue signal.

Space for drawing

1

(ii) The output signal from a computer mouse is digital.

Draw a digital signal.

Space for drawing

1

DO NOT WRITE IN THIS MARGIN

K&U	PS

Marks

13. An electronic system is used to control a lift. When a floor has been selected, two checks are made:

there are no obstructions to the doors;
the lift is not overloaded.

Part of the circuit is shown below.

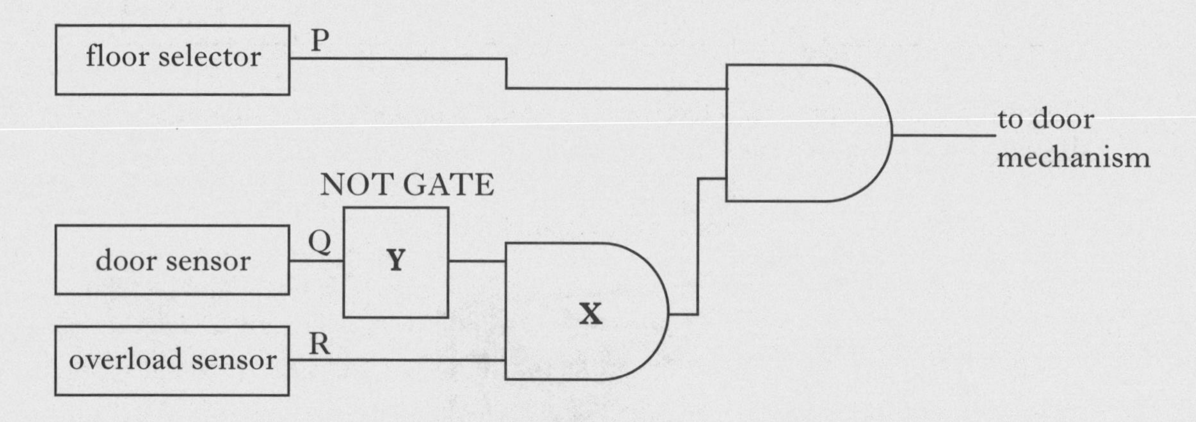

The logic states are as shown for the floor selector, the sensors and the door mechanism.

		logic level
floor selector	not pressed	0
	pressed	1
door sensor	no obstruction	0
	obstruction	1
overload sensor	overloaded	0
	not overloaded	1
door mechanism	doors open	0
	doors closed	1

(*a*) Name logic gate **X**.

.... **1**

DO NOT WRITE IN THIS MARGIN

K&U | PS

Marks

13. (continued)

(*b*) (i) Gate **Y** is a NOT gate.

Draw the symbol for a NOT gate.

Space for symbol

1

(ii) Complete the truth table for a NOT gate.

Input	*Output*
0	
1	

1

(*c*) (i) State the logic levels needed at P, Q and R to close the lift doors.

Logic level at P ....

Logic level at Q ....

Logic level at R 3

(ii) What output device could be used for the door opening and closing mechanism?

.... 1

[Turn over

DO NOT WRITE IN THIS MARGIN

K&U	PS

Marks

14. In a tennis match, the player hits the ball to serve.

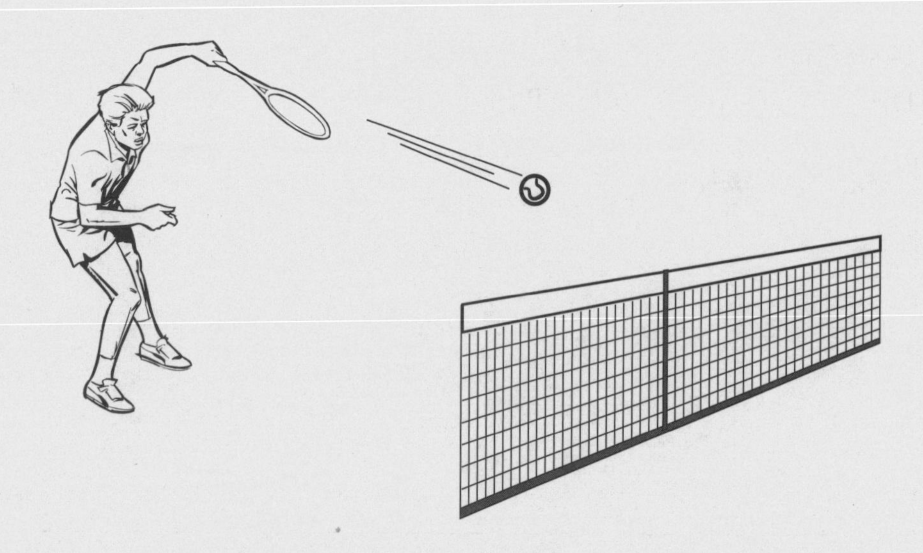

(*a*) The ball travels 24 metres from the server's racquet to the opponent's racquet at an average speed of 40 metres per second.

Calculate the time taken.

Space for working and answer

2

DO NOT WRITE IN THIS MARGIN

K&U	PS

Marks

14. (continued)

(*b*) A graph showing how the speed of the ball changes while in contact with the racquet during the serve is shown.

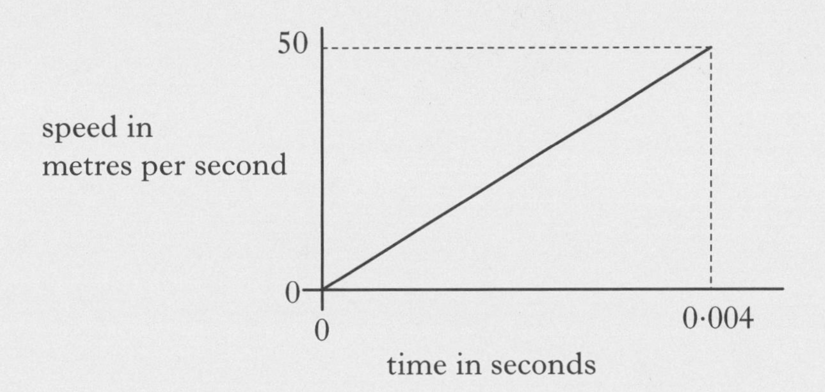

Calculate the acceleration of the ball during the serve.

Space for working and answer

2

(*c*) For a second serve, the server hits the ball with a smaller force.

What effect does this have on the speed of the ball when it leaves the racquet?

.... **1**

[Turn over

DO NOT WRITE IN THIS MARGIN

K&U	PS

Marks

15. A skier takes part in a downhill competition.

(*a*) State **two** ways the skier can reduce friction in order to reach high speeds.

....

.... **2**

(*b*) When the skier reaches the maximum speed of 65 metres per second, this speed is maintained over the rest of the course.

State how the size of the downhill force compares with the size of the frictional force during this part of the course.

.... **1**

(*c*) At the end of the course, the frictional force brings the skier to rest over a horizontal distance of 500 metres. During this distance, the average frictional force is 346 newtons.

Calculate the work done to bring the skier to rest.

Space for working and answer

2

DO NOT WRITE IN THIS MARGIN

K&U	PS

Marks

16. A student carries out an experiment to find out which mug is the best at keeping drinks hot.

Each mug is made from a different material.

plastic

metal

ceramic

The same volume of hot water is added to each mug.

(*a*) Describe how the student could carry out the experiment.

Your description should include:

what apparatus would be used;
what measurements are made;
how you reach a conclusion.

....

....

....

....

....

.... **3**

(*b*) How could the heat lost from the mugs be reduced?

.... **1**

[Turn over

DO NOT WRITE IN THIS MARGIN

K&U | PS

Marks

17. A householder installs a wind turbine electricity generator.

The table gives information about the wind turbine.

Rated power output	1·5 kilowatts
Product life	20 years
Installation cost	£1600

(*a*) In the year 2006, the wind turbine generated electricity for 2000 hours.

Calculate the energy generated in kilowatt-hours during 2006.

Space for working and answer

2

DO NOT WRITE IN THIS MARGIN

K&U | PS

Marks

17. (continued)

(*b*) An electricity supplier charges 8 pence per kilowatt-hour.

Calculate the cost of buying the same amount of electricity as generated by the wind turbine in 2006.

Space for working and answer

2

(*c*) The wind turbine costs £1600 to install. It is used to generate energy for 20 years. Each year it generates the same amount of energy as it did in 2006.

Calculate how much money the householder will save if the turbine is used to generate electricity over this time.

Space for working and answer

2

[Turn over

DO NOT WRITE IN THIS MARGIN

K&U | PS

Marks

18. The diagram below shows a refracting telescope, which is used by astronomers to view distant stars, planets and galaxies.

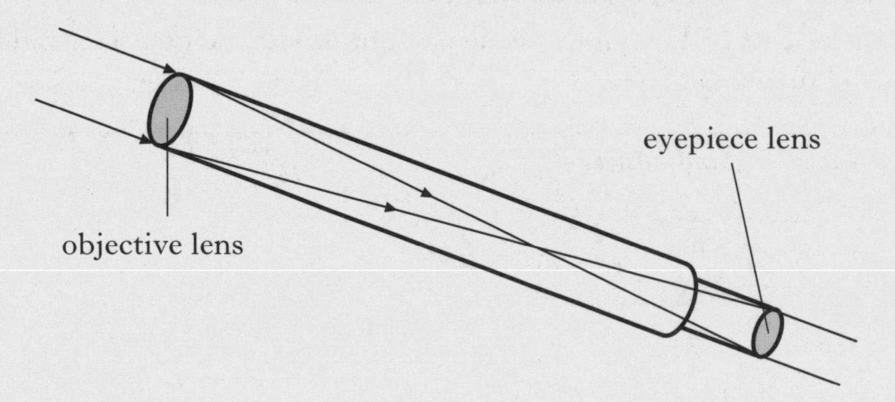

(*a*) (i) Which lens, the objective or the eyepiece, has the longer focal length?

.... **1**

(ii) What is the purpose of the eyepiece lens?

.... **1**

DO NOT WRITE IN THIS MARGIN

K&U PS

Marks

18. (continued)

(*b*) The table gives information about some of the planets in our Solar System.

Planet	*Diameter* (kilometres)	*Distance from Sun* (million kilometres)	*Weight of one kilogram at surface* (newtons)	*Time to go around the Sun once* (years)	*Time for one complete spin* (in Earth days or hours)
Mercury	4800	58	4	0·25	59 days
Venus	12 000	110	9	0·6	243 days
Earth	12 750	150	10	1	24 hours
Mars	7000	228	4	1·9	25 hours
Jupiter	140 000	780	26	12	10 hours
Saturn	120 000	1430	11	30	10 hours
Neptune	50 000	4500	12	165	16 hours

(i) Which planet has the longest day?

.... **1**

(ii) Which planet has the longest orbit?

.... **1**

(iii) On which planet would a 4 kilogram mass have the greatest weight?

.... **1**

(*c*) A meteorite is the name given to an object which enters the Earth's atmosphere from space. When they enter the atmosphere, meteorites heat up.

State the energy change when the meteorite enters the atmosphere.

.... **1**

(*d*) Stars and planets belong to galaxies.

What is a galaxy?

.... **1**

[*END OF QUESTION PAPER*]

DO NOT WRITE IN THIS MARGIN

K&U	PS

YOU MAY USE THE SPACE ON THIS PAGE TO REWRITE ANY ANSWER YOU HAVE DECIDED TO CHANGE IN THE MAIN PART OF THE ANSWER BOOKLET. TAKE CARE TO WRITE IN CAREFULLY THE APPROPRIATE QUESTION NUMBER.

DO NOT WRITE IN THIS MARGIN

K&U	PS

YOU MAY USE THE SPACE ON THIS PAGE TO REWRITE ANY ANSWER YOU HAVE DECIDED TO CHANGE IN THE MAIN PART OF THE ANSWER BOOKLET. TAKE CARE TO WRITE IN CAREFULLY THE APPROPRIATE QUESTION NUMBER.

DO NOT WRITE IN THIS MARGIN

K&U	PS

YOU MAY USE THE SPACE ON THIS PAGE TO REWRITE ANY ANSWER YOU HAVE DECIDED TO CHANGE IN THE MAIN PART OF THE ANSWER BOOKLET. TAKE CARE TO WRITE IN CAREFULLY THE APPROPRIATE QUESTION NUMBER.

DO NOT WRITE IN THIS MARGIN

K&U	PS

YOU MAY USE THE SPACE ON THIS PAGE TO REWRITE ANY ANSWER YOU HAVE DECIDED TO CHANGE IN THE MAIN PART OF THE ANSWER BOOKLET. TAKE CARE TO WRITE IN CAREFULLY THE APPROPRIATE QUESTION NUMBER.

[BLANK PAGE]

[BLANK PAGE]

FOR OFFICIAL USE

G

K & U	PS

Total Marks

3220/401

NATIONAL QUALIFICATIONS 2008

FRIDAY, 23 MAY
9.00 AM – 10.30 AM

PHYSICS
STANDARD GRADE
General Level

Fill in these boxes and read what is printed below.

Full name of centre

Town

Forename(s)

Surname

Date of birth
Day Month Year

Scottish candidate number

Number of seat

Reference may be made to the Physics Data Booklet.

1 All questions should be answered.

2 The questions may be answered in any order but all answers must be written clearly and legibly in this book.

3 For questions 1–5, write down, in the space provided, the letter corresponding to the answer you think is correct. There is only **one** correct answer.

4 For questions 6–20, write your answer where indicated by the question or in the space provided after the question.

5 If you change your mind about your answer you may score it out and replace it in the space provided at the end of the answer book.

6 Before leaving the examination room you must give this book to the invigilator. If you do not, you may lose all the marks for this paper.

SA 3220/401 6/20970

DO NOT WRITE IN THIS MARGIN

K&U PS

Marks

1. When a student whistles a note into a microphone connected to an oscilloscope, the following pattern is displayed.

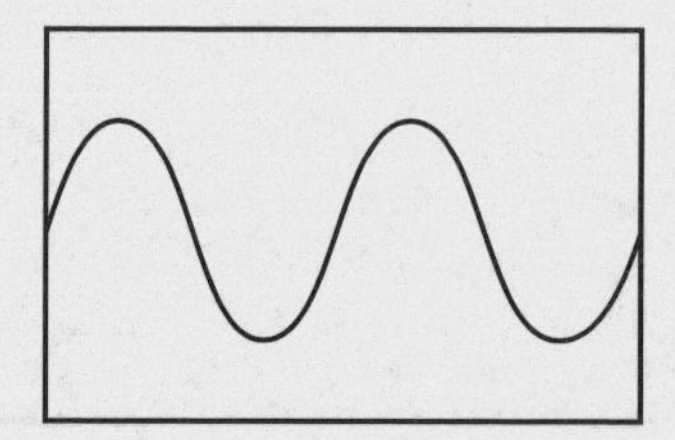

Without changing the oscilloscope controls, another student whistles a quieter note of higher frequency into the microphone. Which of the following shows the pattern which would be displayed on the screen?

A
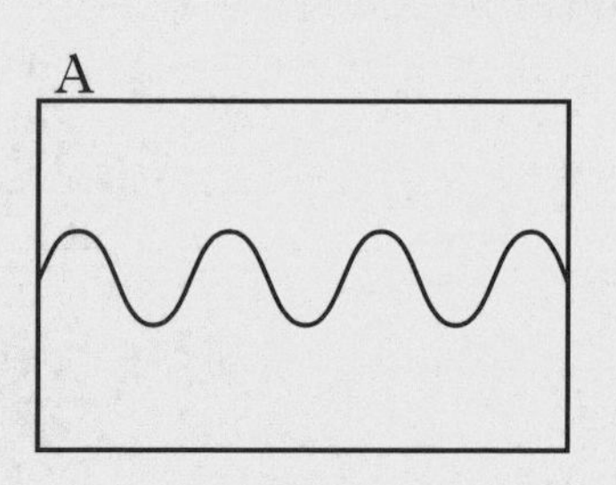

B
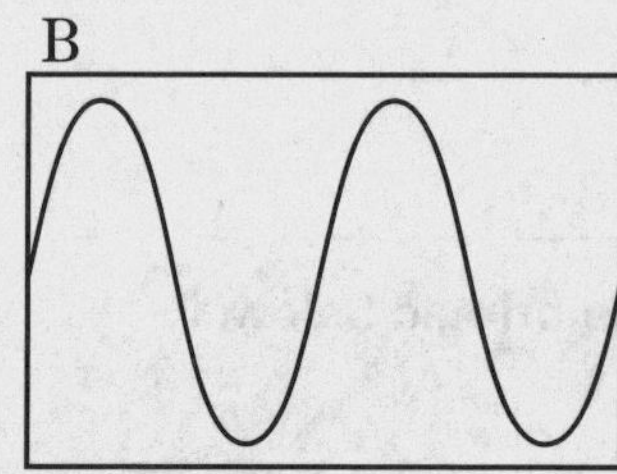

C
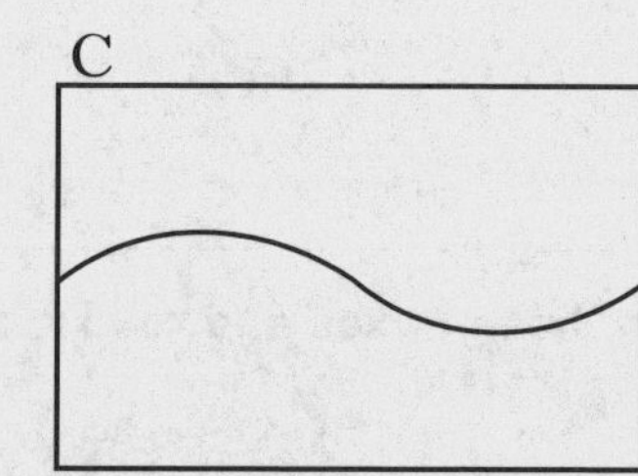

D
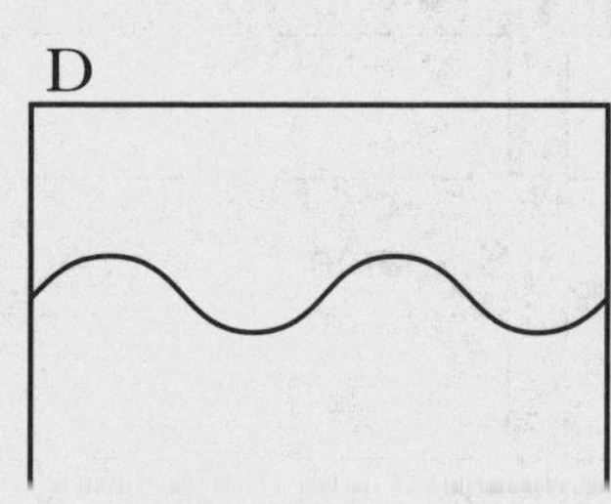

E
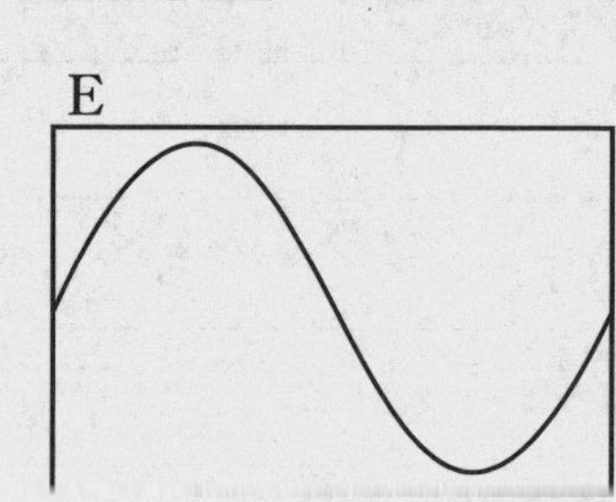

Answer ☐ **1**

2. The weather information satellite NOAA-15 has a period of 99 minutes and an orbital height of 833 kilometres.

The geostationary weather information satellite Meteosat has a period of 1440 minutes and an orbital height of 35 900 kilometres.

Which of the following gives the period of a satellite that has an orbital height of 20 000 kilometres?

A 83 minutes

B 99 minutes

C 720 minutes

D 1440 minutes

E 1750 minutes

Answer ☐ **1**

DO NOT WRITE IN THIS MARGIN

K&U PS

Marks

3. Which row in the table describes the correct configuration for an atom?

	orbiting the nucleus	*inside the nucleus*
A	protons only	electrons and neutrons
B	electrons and protons	neutrons only
C	neutrons and protons	electrons only
D	electrons only	neutrons and protons
E	neutrons only	electrons and protons

Answer ☐ **1**

4. The time taken for light to reach us from the Sun is approximately

A 1 second

B 8 seconds

C 1 minute

D 8 minutes

E 1 hour.

Answer ☐ **1**

5. Two objects are dropped from the same height. Both objects fall freely.

Object X has a mass of 10 kilograms.

Object Y has a mass of 1 kilogram.

Object X accelerates at 10 metres per second per second.

The acceleration of object Y, in metres per second per second, is

A 0·1

B 1·0

C 10

D 100

E 1000.

Answer ☐ **1**

[Turn over

DO NOT WRITE IN THIS MARGIN

K&U	PS

Marks

6. A student is listening to a radio.

(*a*) Complete the passage below using words from the following list.

sound	**amplifier**	**light**	**microphone**	
aerial	**battery**	**tuner**	**decoder**	**electrical**

The .... of a radio receiver detects signals from many different stations and converts them into electrical signals.

The .... selects one particular station from many.

The .... increases the amplitude of these electrical signals.

The energy required to do this is supplied by the ....

The loudspeaker in a radio receiver converts .... energy into .... energy. **3**

DO NOT WRITE IN THIS MARGIN

K&U PS

Marks

6. (continued)

(*b*) Electrical signals are displayed as waves on an oscilloscope.

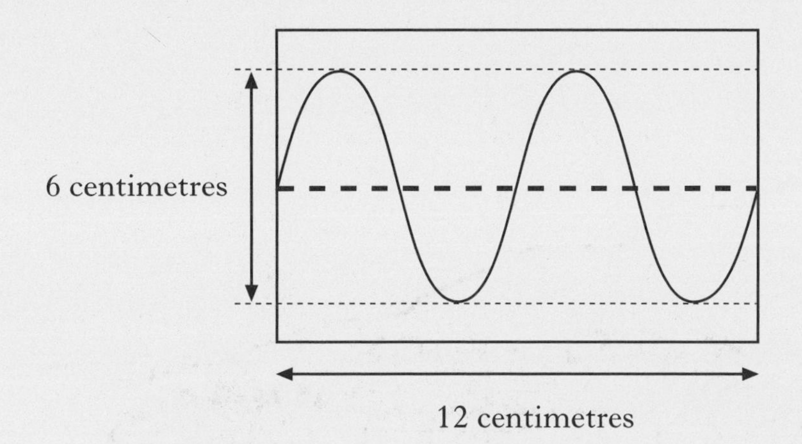

(i) Calculate the wavelength of the waves.

Space for working and answer

1

(ii) Calculate the amplitude of the waves.

Space for working and answer

1

[Turn over

DO NOT WRITE IN THIS MARGIN

K&U | PS

Marks

7. A football match is being broadcast live from Dundee. Signals from the football stadium are transmitted to a television studio in Glasgow via a relay station on top of a nearby hill.

At the relay station, a curved reflector is placed behind a detector of the television signals.

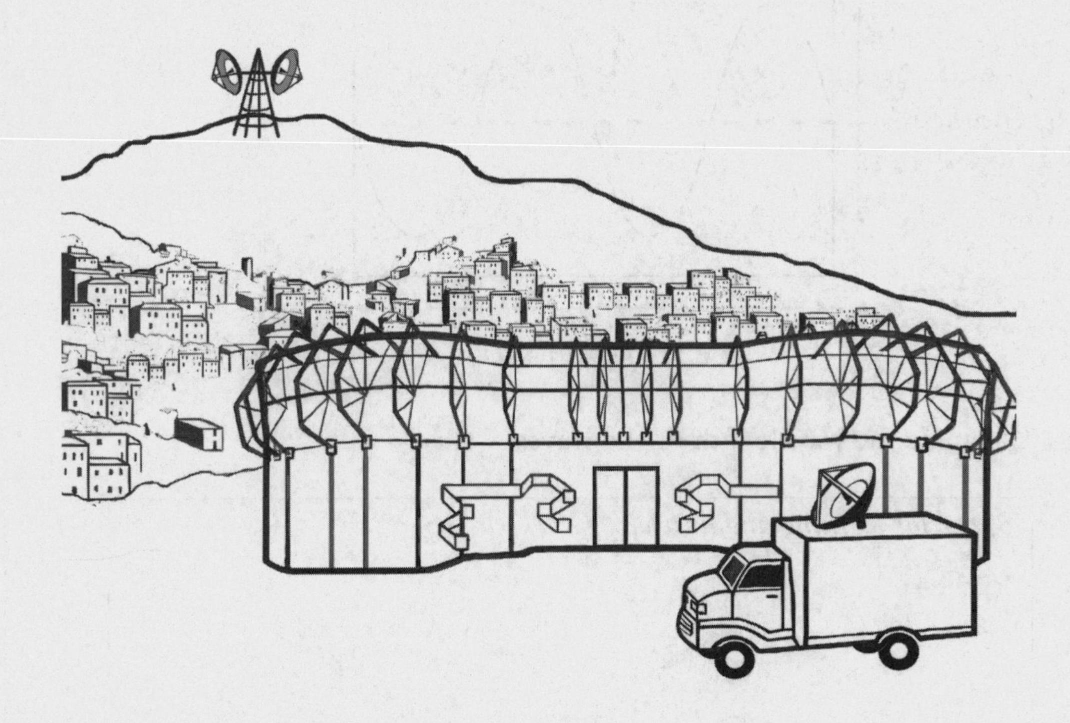

(*a*) (i) State the purpose of the curved reflector.

.... **1**

(ii) Complete the diagram below to show the effect of the curved reflector on the signal at the relay station.

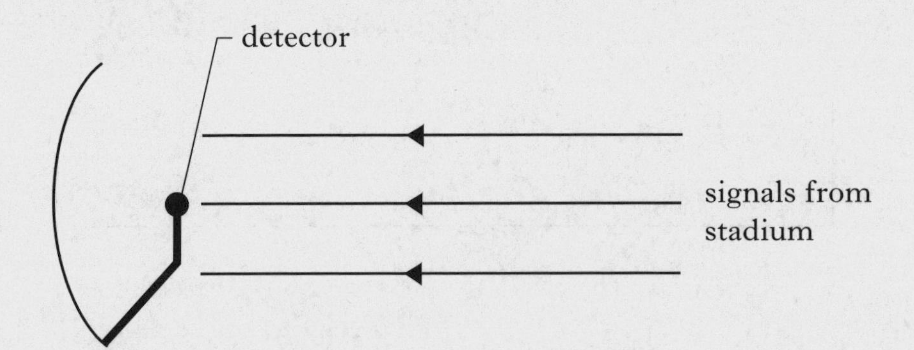

2

DO NOT WRITE IN THIS MARGIN

K&U	PS

Marks

7. (continued)

(*b*) During the match, strong winds cause the reflector to move to a new position as shown.

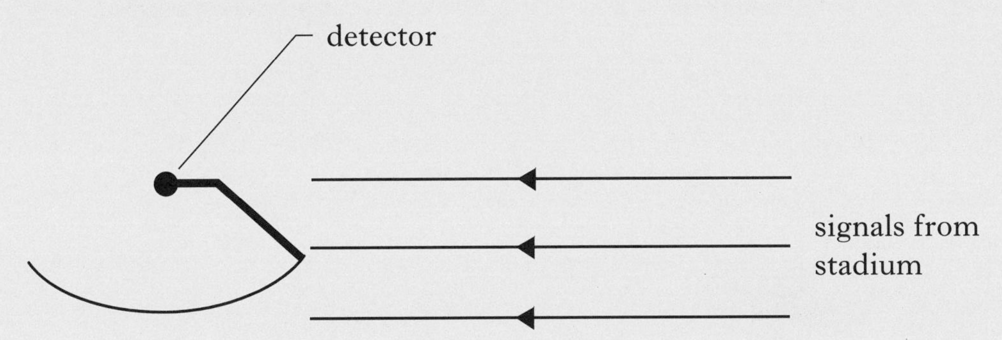

State the effect this has on the signal received at the detector.

.... **1**

[Turn over

DO NOT WRITE IN THIS MARGIN

K&U	PS

Marks

8. Two household electrical appliances, a 1500 watt electric iron and a 300 watt uplighter lamp, are shown below.

electric iron

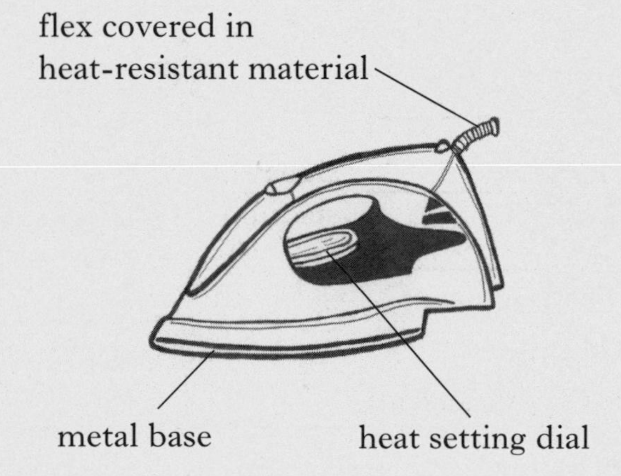

uplighter lamp

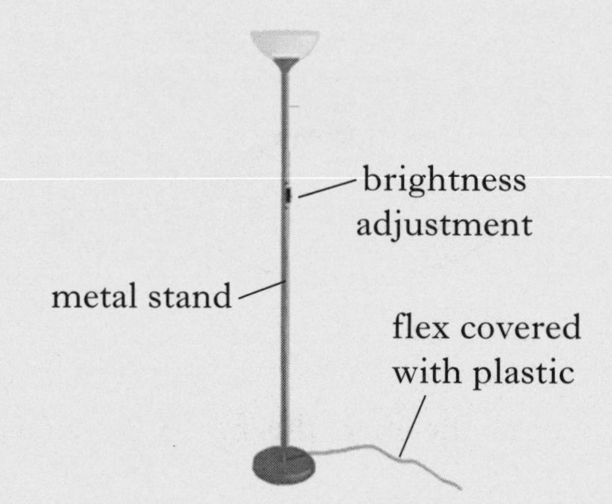

(*a*) The brightness of the uplighter lamp can be changed.

State an electrical component that could be used to change the brightness of the uplighter lamp.

.... **1**

(*b*) Explain why the flex for the iron is covered with a heat-resistant material.

.... **1**

DO NOT WRITE IN THIS MARGIN

K&U PS

Marks

8. (continued)

(*c*) A cross-section of the flex for each appliance is shown.

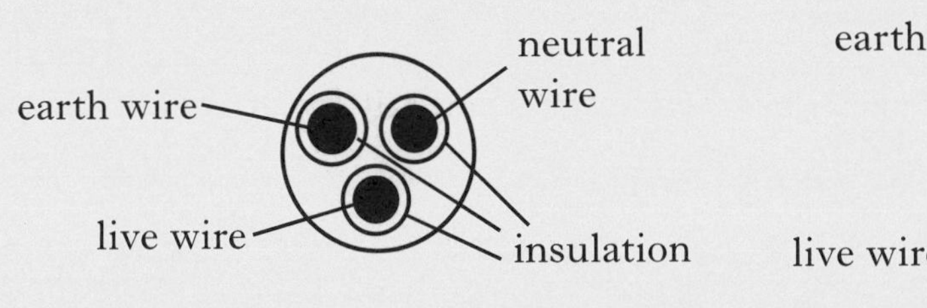

(i) State the colour of the insulation on the live wire.

.... **1**

(ii) State the purpose of the earth wire.

.... **1**

(iii) Explain why the wires in the flex for the electric iron are thicker than those for the uplighter lamp.

.... **1**

[Turn over

DO NOT WRITE IN THIS MARGIN

K&U	PS

Marks

9. Two identical lamps are connected to a 6·0 volt battery as shown in circuit 1.

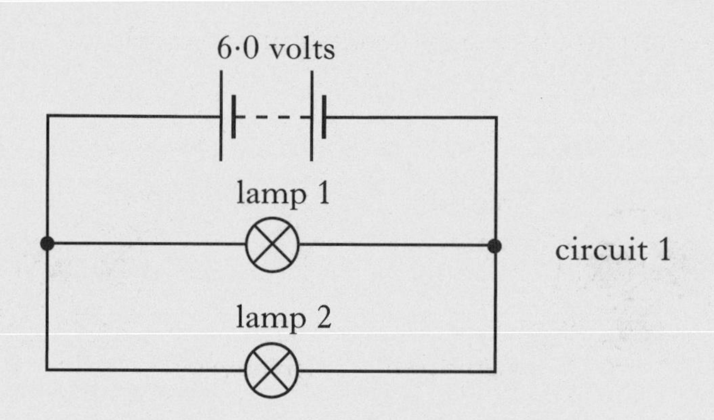

(*a*) The battery supplies a current of 0·40 ampere to the circuit.

Complete the following table to show the current in each lamp and the voltage across each lamp.

	Lamp 1	**Lamp 2**
Current (amperes)		
Voltage (volts)		

2

(*b*) The two lamps are now connected as shown in circuit 2.

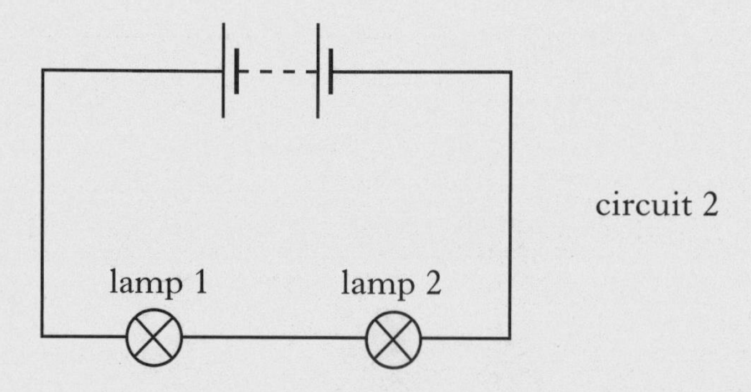

State the voltage of the battery required to light the lamps with the same brightness as in circuit 1.

.... **1**

(*c*) In which of the two circuits, circuit 1 or circuit 2, would lamp 2 still be on when lamp 1 is removed?

.... **1**

DO NOT WRITE IN THIS MARGIN

K&U | PS

Marks

10. (*a*) A drummer in a rock band is exposed to sound levels of up to 110 decibels.

Explain why ear protectors are used to reduce the sound level experienced by the drummer.

.... **1**

(*b*) A medical researcher is measuring the upper range of hearing of people in different age groups.

The bar graph shows the frequencies of sound detected by these people.

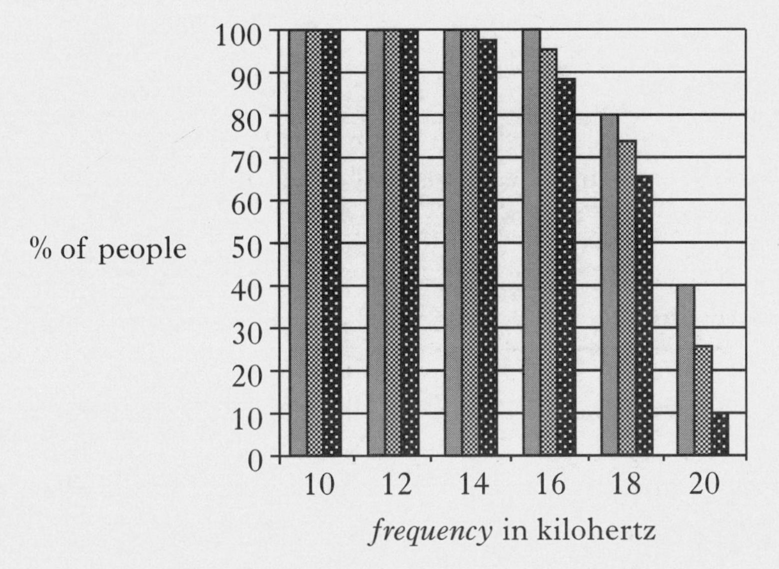

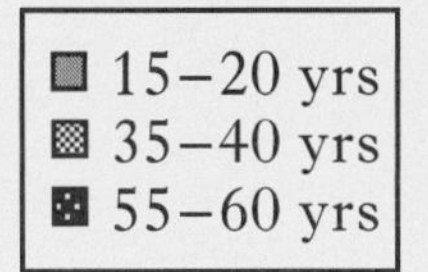

(i) State **two** conclusions which can be made from this bar graph about the hearing of different age groups.

....

.... **2**

(ii) What name is given to sound frequencies greater than 20 kilohertz?

.... **1**

DO NOT WRITE IN THIS MARGIN

K&U | PS

Marks

11. (*a*) A thermistor is connected to a 6·0 volt supply in circuit 1. The table gives some information about the thermistor.

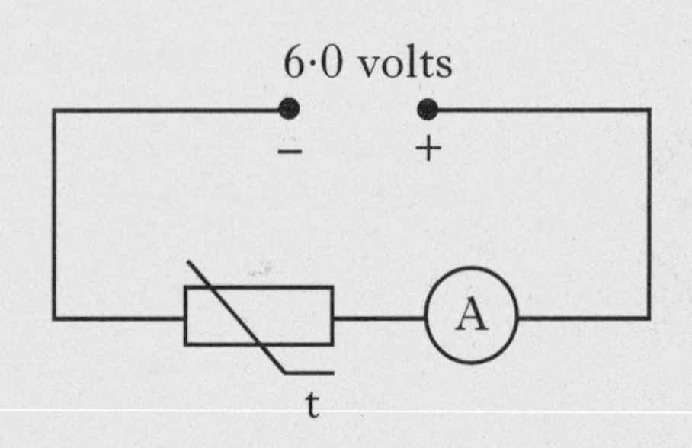

circuit 1

temperature (degrees Celsius)	*resistance* (ohms)
20	1000
30	600
40	400

Calculate the reading on the ammeter when the thermistor is placed in a beaker of water at 40 degrees celsius.

Space for working and answer

3

(*b*) The thermistor is now connected as shown in circuit 2 and placed in a tropical fish tank. The circuit provides a warning when the temperature of the water in the tank becomes too low.

circuit 2

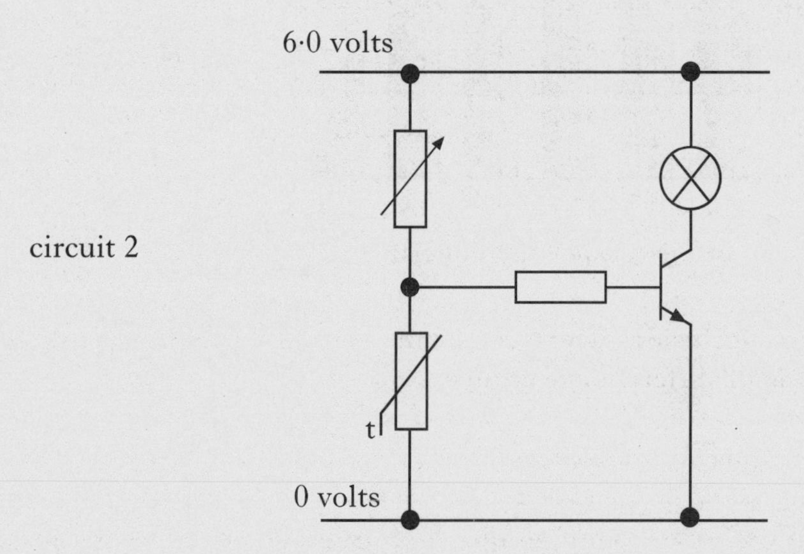

(i) What is the purpose of the transistor in circuit 2?

.... 1

DO NOT WRITE IN THIS MARGIN

K&U	PS

Marks

11. (*b*) (continued)

(ii) The same components are used to construct circuit 3.

circuit 3

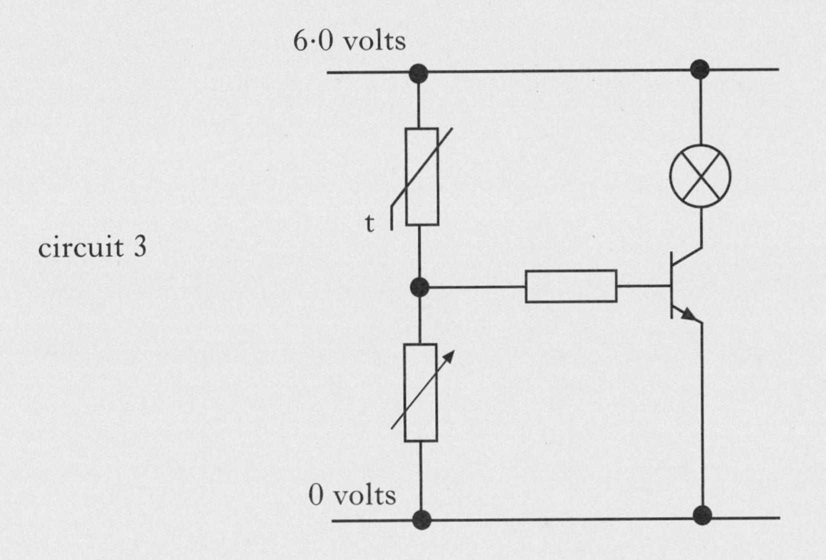

State how the operation of circuit 3 differs from the operation of circuit 2.

.... **1**

[Turn over

DO NOT WRITE IN THIS MARGIN

K&U PS

Marks

12. (*a*) A nurse uses a clinical thermometer to measure the body temperature of a patient. The temperature of the patient is 39 degrees celsius.

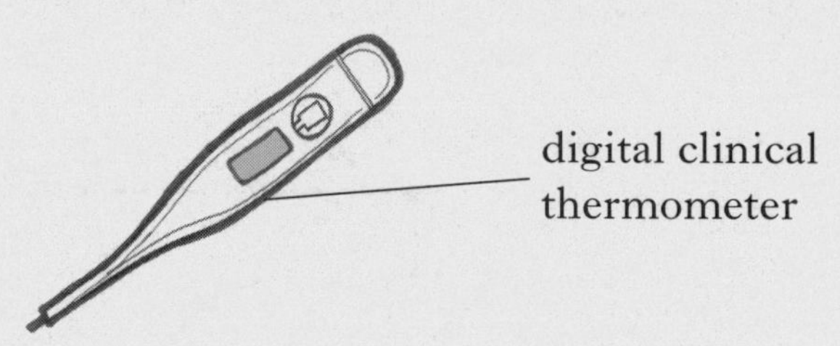

(i) Give **two** reasons why a clinical thermometer is used instead of an ordinary thermometer when measuring the body temperature of the patient.

....

.... **2**

(ii) Why does the nurse conclude that the patient is unwell?

.... **1**

(*b*) Radioactive sources are used in the treatment of many illnesses. The table below gives some properties of three radioactive sources used in medicine.

Name of Source	*Type of Source*	*Half-life of Source*
Radium – 226	Alpha	1600 years
Iodine – 131	Beta	8 days
Technetium – 99	Gamma	6 hours

(i) One type of treatment requires a source that produces high ionisation.

Which source should be used?

.... **1**

(ii) Which source would be most suitable for use in diagnostic tests where a tracer is injected into the body?

.... **1**

(iii) Which source should not be stored in an aluminium box for safety reasons?

.... **1**

DO NOT WRITE IN THIS MARGIN

K&U | PS

Marks

13. An electronic system is designed to count the number of vehicles that enter a car park.

When a vehicle enters the car park it cuts through a beam of light and a sensor circuit produces a digital pulse. The number of pulses produced by the sensor circuit is then counted and decoded before being displayed. The display consists of a number of illuminated sections.

A diagram for part of this system is shown.

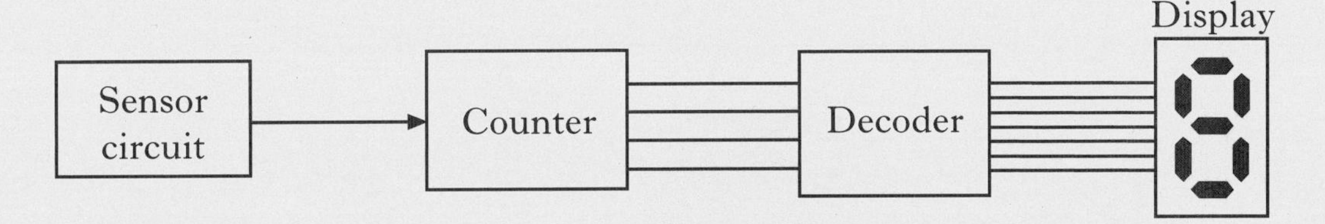

(*a*) (i) Select a suitable device **from the list below** to be used as an input for the sensor circuit.

LDR **thermistor** **microphone** **capacitor**

.... **1**

(ii) Complete the sentence below by circling the correct answer.

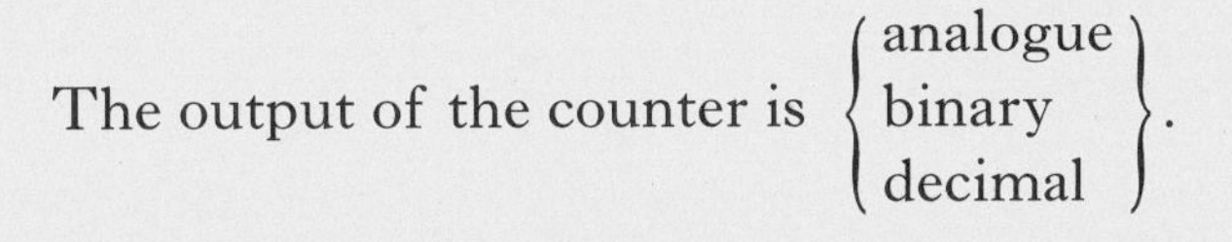

1

(iii) Name the device used to display the number of vehicles that enter the car park.

.... **1**

(*b*) The counter is reset to zero. Over a period of time, the sensor circuit then produces the following signal.

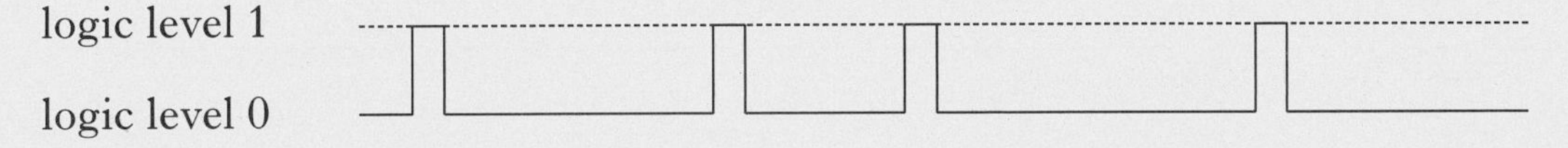

On the diagram of the display below, shade in the sections that should be illuminated to show the number of vehicles that have entered the car park during this time.

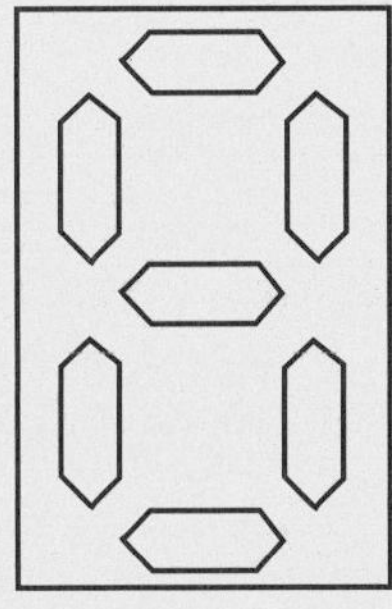

1

DO NOT WRITE IN THIS MARGIN

K&U	PS

Marks

14. A walker wears a pedometer. A pedometer is an instrument that measures the distance walked by counting the number of steps taken. The walker measures the distance of one step as 0·8 metres, and enters it into the pedometer.

(*a*) The walker completes 9000 steps during a walk.

Calculate the distance travelled.

Space for working and answer

1

(*b*) The walker completes this walk in 80 minutes.

What is the average speed of the walker in **metres per second**?

Space for working and answer

2

(*c*) Give a reason why the distance measured by the pedometer may not be accurate.

.... **1**

DO NOT WRITE IN THIS MARGIN

K&U	PS

Marks

15. A piano of mass 250 kilograms is pushed up a ramp into a van by applying a constant force of 600 newtons as shown.

The ramp is 3 metres long and the van floor is 0·75 metres above the ground.

(*a*) (i) Calculate the weight of the piano.

Space for working and answer

2

(ii) What is the minimum force required to lift the piano vertically into the van?

.... **1**

(*b*) Calculate the work done pushing the piano up the ramp.

Space for working and answer

2

(*c*) How can the force required to push the piano up the ramp be reduced?

.... **1**

[Turn over

DO NOT WRITE IN THIS MARGIN

K&U	PS

Marks

16. A traffic information sign is located in a remote area.

The sign is supplied with energy by both a panel of solar cells and a wind generator. The panel of solar cells and the wind generator are connected to a rechargeable battery.

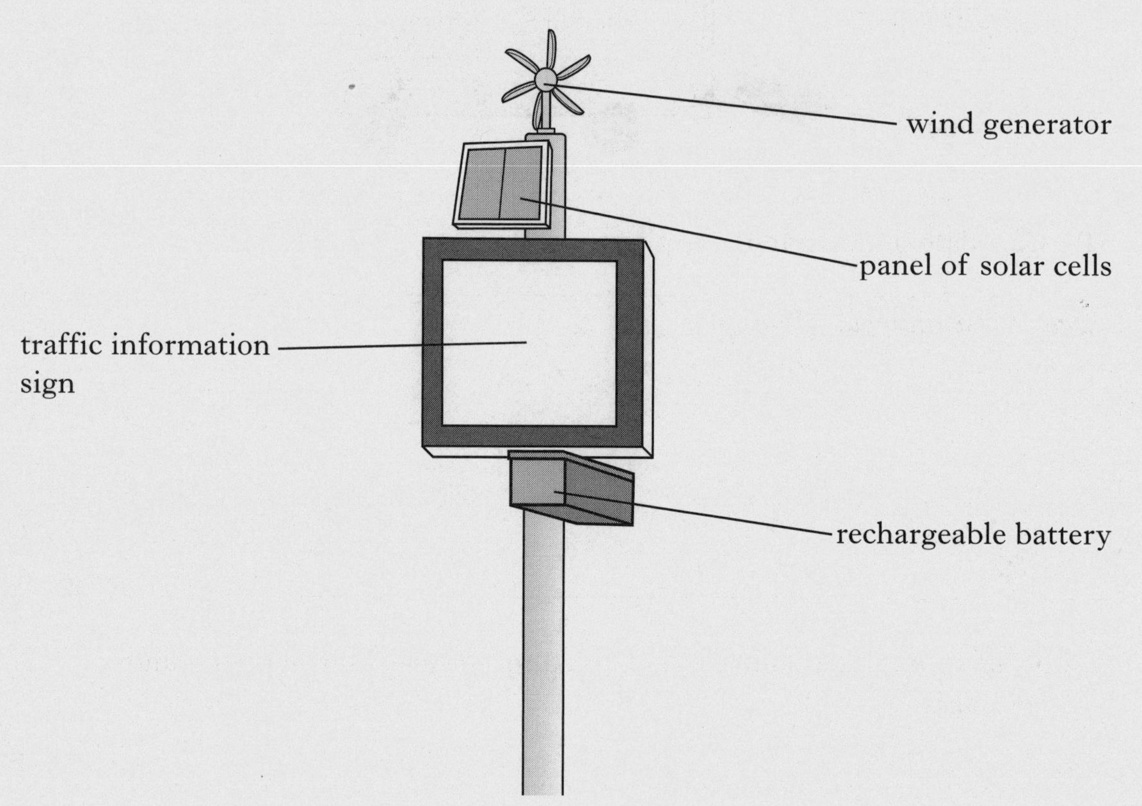

(*a*) One square metre of solar cells can generate up to 80 watts.
The panel of solar cells has an area of 0·4 square metres.

(i) State the energy change that takes place in the solar cells.

.... **1**

(ii) Calculate the maximum power produced by the panel of solar cells.

Space for working and answer

1

DO NOT WRITE IN THIS MARGIN

K&U PS

Marks

16. (continued)

(*b*) The following table shows the power produced by the wind generator at different wind speeds.

wind speed (metres per second)	*power output of wind generator* (watts)
2	8
4	16
6	
8	32
10	40

(i) Suggest the power produced when the wind speed is 6 metres per second.

.... **1**

(ii) At a wind speed of 10 metres per second the voltage produced by the wind generator is 16 volts.

Calculate the current produced by the wind generator.

Space for working and answer

2

(*c*) Explain why a rechargeable battery is also required to supply energy to the traffic information sign.

.... **1**

[Turn over

DO NOT WRITE IN THIS MARGIN

K&U	PS

Marks

17. (*a*) A digital camera contains a rechargeable battery. The battery requires a voltage of 5·75 volts to be recharged. The battery is recharged using a transformer connected to the mains supply. The transformer is used to step down the 230 volt a.c. mains supply to 5·75 volts.

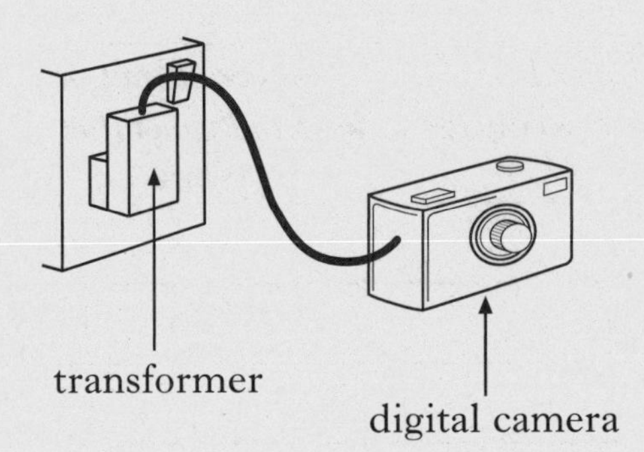

The transformer has 2000 turns on the primary coil.

(i) Calculate the number of turns on the secondary coil.

Space for working and answer

2

(ii) Give **one** reason why a transformer cannot be used to charge the camera battery from a 12 volt d.c. car battery.

.... **1**

(*b*) Complete the following passage.

In the National Grid, ... transformers are used to increase the 25 000 volts from a power station to 132 000 volts for transmission.

This reduces .... in the transmission lines.

The voltage is then decreased to 11 000 volts for industry and 230 volts for domestic use using .... transformers. **3**

DO NOT WRITE IN THIS MARGIN

K&U PS

Marks

18. A coolant pack is used to treat an injured player at a hockey match.

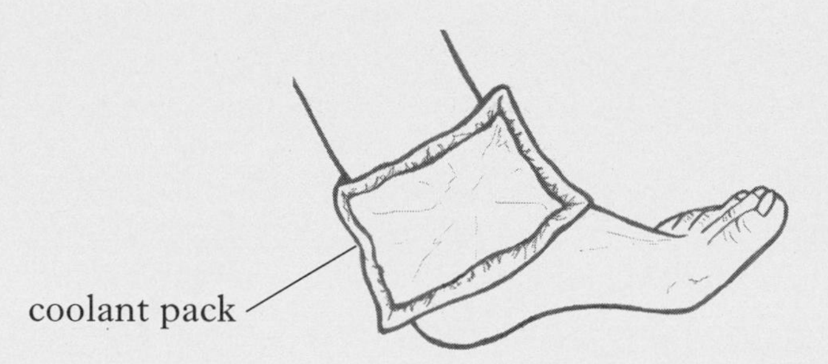

Before use the coolant pack is stored in a refrigerator at 2 degrees celsius.

The coolant inside the pack changes state from liquid to solid.

The coolant has a melting point of 7 degrees celsius and a mass of 0·5 kilograms.

The coolant pack is removed from the refrigerator and placed on the injured ankle of a player.

(*a*) (i) Calculate the energy required to raise the temperature of the coolant pack from 2 degrees celsius to its melting point.

(specific heat capacity of coolant = 2100 joules per kilogram per degree celsius)

Space for working and answer

3

(ii) Where does most of the energy required to raise the temperature of the coolant pack come from?

.... **1**

(*b*) Having reached its melting point the coolant pack then remains at the same temperature for 15 minutes.

What is happening to the coolant during this time?

.... **1**

(*c*) One of the other players suggests insulating the coolant pack and ankle with a towel.

Why should this be done?

.... **1**

DO NOT WRITE IN THIS MARGIN

K&U	PS

Marks

19. Read the following passage about a space mission to the moons of Jupiter.

The spacecraft will use a new kind of engine called an ion drive. The ion drive will propel the spacecraft away from Earth on its journey to the moons of Jupiter, although for much of the journey the engine will be switched off.

The spacecraft will first visit the moon Callisto.

Callisto is only slightly smaller than the planet Mercury. Next, the spacecraft will visit Ganymede, the largest moon in the Solar System, before travelling on to Europa.

The radiation around Europa is so intense that the spacecraft will not be able to operate for long before becoming damaged beyond repair.

The spacecraft will eventually burn up in the atmosphere of Jupiter.

(*a*) (i) Name one object, **mentioned in the passage**, which orbits a planet.

.... **1**

(ii) State what is meant by the term Solar System.

.... **1**

(*b*) (i) The ion drive engine exerts a backward force on small particles called ions.

Explain how the ion drive engine is propelled forwards.

.... **1**

(ii) The mass of the spacecraft is 1200 kilograms and the thrust produced by the engine is 3 newtons.

Calculate the maximum acceleration produced by the ion drive engine.

Space for working and answer

2

(*c*) State why the ion drive engine need not be kept on for most of the journey from Earth to Jupiter.

.... **1**

DO NOT WRITE IN THIS MARGIN

K&U PS

Marks

20. (*a*) A ray of green light strikes a triangular prism as shown.

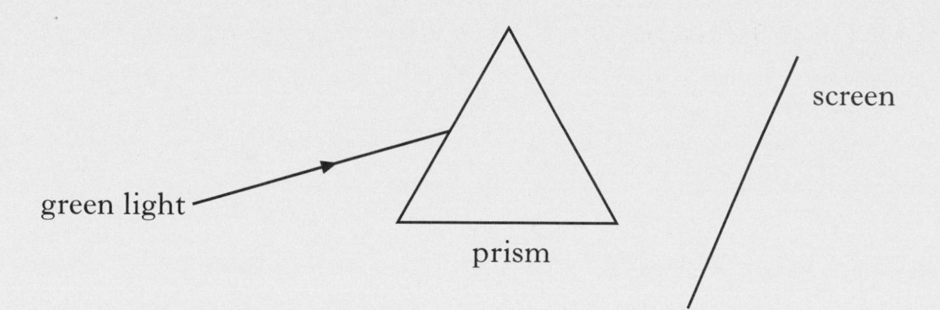

(i) Complete the diagram to show the path of the ray of green light as it passes through the prism and on to the screen. **1**

(ii) The green light is now replaced by white light.

Describe what is now observed on the screen.

.... **1**

(iii) State **one** colour which has a longer wavelength than green light.

.... **1**

(*b*) Light from a star produces a line spectrum.

What information is obtained about the star from this spectrum?

.... **1**

[*END OF QUESTION PAPER*]

DO NOT WRITE IN THIS MARGIN

K&U	PS

YOU MAY USE THE SPACE ON THIS PAGE TO REWRITE ANY ANSWER YOU HAVE DECIDED TO CHANGE IN THE MAIN PART OF THE ANSWER BOOKLET. TAKE CARE TO WRITE IN CAREFULLY THE APPROPRIATE QUESTION NUMBER.

DO NOT WRITE IN THIS MARGIN

K&U	PS

YOU MAY USE THE SPACE ON THIS PAGE TO REWRITE ANY ANSWER YOU HAVE DECIDED TO CHANGE IN THE MAIN PART OF THE ANSWER BOOKLET. TAKE CARE TO WRITE IN CAREFULLY THE APPROPRIATE QUESTION NUMBER.

DO NOT WRITE IN THIS MARGIN

K&U	PS

YOU MAY USE THE SPACE ON THIS PAGE TO REWRITE ANY ANSWER YOU HAVE DECIDED TO CHANGE IN THE MAIN PART OF THE ANSWER BOOKLET. TAKE CARE TO WRITE IN CAREFULLY THE APPROPRIATE QUESTION NUMBER.

DO NOT WRITE IN THIS MARGIN

K&U	PS

YOU MAY USE THE SPACE ON THIS PAGE TO REWRITE ANY ANSWER YOU HAVE DECIDED TO CHANGE IN THE MAIN PART OF THE ANSWER BOOKLET. TAKE CARE TO WRITE IN CAREFULLY THE APPROPRIATE QUESTION NUMBER.

[BLANK PAGE]

[BLANK PAGE]

[BLANK PAGE]